CHIEFSHIP AND COSMOLOGY

African Systems of Thought

General Editors
Charles S. Bird
Ivan Karp

Contributing Editors
James Fernandez
Luc de Heusch
John Middleton
Victor Turner
Roy Willis

CHIEFSHIP
AND
COSMOLOGY

AN HISTORICAL STUDY
OF POLITICAL COMPETITION

Randall M. Packard

INDIANA UNIVERSITY PRESS
Bloomington

Copyright © 1981 by Randall M. Packard

Library of Congress Cataloging in Publication Data

Packard, Randall M., 1945–
Chiefship and cosmology.

(African systems of thought)
Includes bibliographical references and index.
1. Bashi (African people)—Kings and rulers.
2. Bashi (African people)—Politics and government.
I. Title. II.Series.
DT650.B.366P3   967.5'17   81-47013
ISBN 0-253-30831-3   AACR2
1 2 3 4 5 85 84 83 82 81

CONTENTS

MAPS

TABLES

ORTHOGRAPHY

There is presently no official orthography for Kinande or for Kishu, the dialect of Kinande spoken by the inhabitants of the Bashu chiefdoms. There is, however, an extensive Kinande-English Dictionary compiled by Pauline Fraas (1961), primarily from Baswaga informants. I have for the most part followed Fraas's orthography, with two notable exceptions. First, Fraas represents the bilabial fricative, found in many Kinande words, as 'b,' regardless of the following vowel. In Kishu this phoneme is closer to 'b' before 'a' but closer to 'v' before 'u.' Thus I have written *baloyi,* 'sorcerers,' but *vuloyi,* 'sorcery.' Second, Fraas represents the velar fricative of Kinande as 'g.' I have instead used 'gh' to represent this phoneme (thus *vighala, baghula* and *akaghenda*), a choice that follows the general practice of literate Bashu and avoids confusion with the hard 'g' of English. Finally, for simplicity, I have dropped the pre-prefixes 'o,' 'a,' 'e,' and 'i' from all Kinande words, except where pronunciation is impeded by this practice, and the prefix 'eri-' from all infinitives.

ACKNOWLEDGMENTS

Shortly after I began this study, I found myself making a mental list of the people who had assisted me in its preparation. The extent to which I am indebted to these people and their importance to the completion of this study can perhaps only be understood by those who have undertaken similar projects and incurred similar debts of gratitude. Nonetheless, I wish to take this opportunity to acknowledge their assistance.

First I wish to thank the Fulbright-Hays Dissertation Year Abroad Program and the University of Wisconsin–Madison for funding my research in Belgium and Zaire. Additional funds for the preparation of the present manuscript were generously provided by the Tufts University Faculty Research Fund.

Second, I wish to thank Drs. Steven Feierman and Jan Vansina, who worked closely with me during the course of my field work and during the writing of the Ph.D. dissertation out of which the present study grew. Dr. Feierman's own work on the political history and culture of the Shambaa of Tanzania provided me with important insights into the relationship between ideology and action in African politics. A number of other scholars at the University of Wisconsin, including Drs. Philip Curtin, Aiden Southall, Patrick Bennett, David Henige, Richard Sigwalt, Ellie Sosne, and Ephraim Kamuhangire, gave generously of their time and advice while I was preparing to go into the field. I am particularly indebted to Dr. Ivan Karp of Indiana University. From our many long and fruitful discussions I have gained numerous insights into the nature and meaning of myth, symbol, and ritual within the African experience, insights that have contributed in no small way to my understanding of the politics of ritual chiefship among the Bashu.

During the course of my research, I incurred additional debts of gratitude. In Belgium, I was greatly assisted by the staff of the Musée Royal de l'Afrique Centrale in Tervuren, who allowed me the use of their research and living facilities. I especially want to thank Dr. Luc Cahen, former director of the museum, for his assistance in obtaining research clearance for my work in Zaire, as well as for his hospitality while at Tervuren; Dr. Marcel d'Hertefelt, chef de section ethnographique, for his advice and comments regarding research in Zaire; and Mme. Denise Caneel, whose many kindnesses made my stay in Belgium a pleasant one.

I am also indebted to the Government of Zaire for having allowed me to carry out my research among the Banande and to consult archival sources within the country, and to l'Institut pour la Recherche Scientifique en Afrique Centrale, and its former director Dr. Ntika-Nkumu, for helping me prepare for my field work in Bunande. Drs. Catherine and David Newbury, former research associates at I.R.S.A.C., were similarly helpful during my stay there.

Above all, I wish to thank the hundreds of men, women, and children among the Banande who helped me during my research, answering my questions, guiding me along the mountain paths of the Mitumbas and Ruwenzoris, and welcoming me into their homes. I especially wish to thank Kataka Mbungu, my colleague and field assistant among the Banande, and Mombokani Wasokundi, my guide and mentor in Bashu history. In a very real sense, this study would never have been completed without their dedication and companionship.

During my field work, I also benefitted from long talks with Père Lieven Bergmans, whose knowledge and study of Nande history and culture were of great value to my research, as well as with John Hart, an American researcher working among the Bambuti of the Ituri Forest. I also want to thank Paul and Fay Hurlburt, Larry and Sandy Mahlum, and the Assumptionist Catholic Fathers of Butembo, Kyondo, Paida, Lubango, Mutwanga, Bunyuka, and Luofu for their hospitality while I was in Zaire.

At various stages in the preparation of this study I benefitted from comments and suggestions by Drs. Steven and Jane Baier, Frederick Cooper, Bertin Webster, John Rowe, Jim Freedman, Crawford Young, and Michael Kenny, and from discussions with my colleagues and students at Tufts. I would also like to thank Dr. David Cohen for the meticulous care he took in reading and commenting on this manuscript for the Press.

Finally, I wish to thank my wife, Carolyn—who since the beginning of our marriage has had to share me with the Bashu—for her patience, encouragement, and love; and my parents, to whom this book is dedicated, for their pride and support over the many years of preparation that have led to the publication of this study.

INTRODUCTION

THE SETTING

This study began among the Bashu of eastern Zaire in the spring of 1974 with an unusually long rainy season and the threat of famine. The Bashu are one of several major clans that make up the Banande peoples, who occupy the Mitumba Mountains to the north and west of Lake Edward,[1] in what is now the Kivu Region of Zaire. They are related historically, linguistically, and culturally to the Bakonjo of western Uganda.[2]

While Bashu are found throughout Bunande, their main areas of settlement are located in the Isale region of the Mitumbas along the western wall of the upper Semliki Valley, and in the regions of Malio and Maseki immediately adjacent to Isale. Here the Bashu form the dominant demographic and political group in several closely related chiefdoms. Other Nande clans that live within the Bashu chiefdoms presently identify themselves with the dominant Bashu and often refer to themselves as Bashu.[3] The term 'Bashu,' therefore, denotes the people living within the Bashu chiefdoms as well as members of the Bashu clan. Unless otherwise noted, the term is used in its broader sense in the present study. To the west of the Bashu chiefdoms, members of the Baswaga clan enjoy a similar position of dominance, while further south, along the western shore of Lake Edward, the Bamate and Batangi are the politically dominant clans.

The Bashu are agriculturalists, growing crops of plantain bananas, eleusine, sorghum, beans, cassava, and sweet potatoes, along with more recently introduced cash crops of coffee and wheat. They also raise goats, chickens, and a few head of cattle. In addition, the Bashu have a long history of involvement in regional commerce going back to their early participation in the salt trade that centered on Lake Katwe in Uganda. In 1974 the Bashu along with other Nande groups dominated commercial activity over much of the Kivu Region.

Politically, the Bashu chiefdoms are ruled by a family of chiefs who claim to be related to the Babito rulers of Bunyoro and Toro. The ancestors of these chiefs are said to have settled in the Mitumbas during the early years of the nineteenth century. The Bashu chiefdoms were consolidated into a collectivity and incorporated into the Belgian Congo in 1923, and into the independent Congo Republic, now Zaire, in 1960.

I arrived among the Bashu in February of 1974 and settled in the village of Kahondo, which is located in Isale near the present administrative post of Vuhovi. At the time of my arrival the people of Kahondo were preparing their fields for planting beans, a primary source of protein in the Bashu diet. The men cleared away the bush and broke up the soil. Once this was completed, the women prepared the seed mounds and planted the seeds in them. The people of Kahondo then waited for the short rains of March and April, which would germinate and nourish their crops.[4]

The rains came on time during the second week of March, in a mixture of thunderstorms and long soaking rains, and by the first week in May the people of Kahondo began looking for the rains to end and for the return of drier weather, which would bring the beans to maturity. But the rains continued to fall, and by the end of May, Bashu men and women began to express concern over the well-being of their crops.

When the rains still had not ended by the middle of June, concern gave way to active discussions of what was causing the rains to continue, and, more importantly, what could be done to stop them. Here and there 'diviners of the land' were consulted, offerings were made to ancestors and particular earth spirits, and groups of elders began visiting local rainmakers, who had the power to stop the rains as well as to start them. But the rains continued to fall and people began speaking of *nzala*, the famine that would result from the loss of their crops.

The rains finally came to an end during the first week in July. But the damage had already been done. One-half to three-quarters of the bean crop had rotted on the bush. Much of that which remained had been picked prematurely to avoid total destruction and was consequently of limited nutritional value.

What had caused the rains to continue beyond their normal period? The question occurred again and again in daily conversations held in the village meeting house, around the cooking pots at night, and in the fields during the day as the Bashu prepared their land for the planting of sorghum and eleusine. Various answers were given. A few adults who were active Christians attributed the rains to divine providence, and some young men who had attended mission schools offered an explanation based on their understanding of meteorological science. But the older men and women, the elders, knew better. For them, the prolonged rains were the direct result of the failure of the people of Isale to agree on, and invest, a central ritual chief, a *mwami w'embita*.

The last properly invested chief had died in 1950 and the loss of the bean crop was but one of a series of misfortunes that had disrupted the region since his death. Everything from the invasion of the Simba rebels in 1964 to the locusts that threatened to descend upon Isale in December of 1974 was symptomatic of this failure to invest a proper *mwami*.

THE PROBLEM

The Bashu elders' explanation for the extended rains of 1974 provides an important key to understanding Bashu chiefship and politics. From an external viewpoint, Bashu chiefs can be seen as political leaders. They have the right to claim and receive tribute, to determine the distribution and use of land, and to levy fines for the transgression of certain social and religious prohibitions. In addition, they often act like politicians. They acquire their position by competing with existing chiefs or by engaging in succession struggles with rival kinsmen, and they strengthen their subsequent authority by competing with neighboring chiefs or potential rivals within their own chiefdoms. In other words, they participate in competitive political activities that have been noted in a number of other African states and carefully examined in the processual studies of Southall (1953), Fallers (1965), Gluckman (1954), Goody (1966), and Richards (1960 and 1961).

Yet to view Bashu chiefs simply as political leaders or politicians is to ignore how the Bashu themselves conceive of them and of the competitive actions that surround the office of chiefship. Bashu conceptions of chiefship and politics are reflected in the Bashu elders' explanation for the rains of 1974 and flow from their understanding of the world in which they live and of the forces that shape that world, from what Rappaport (1968: 327–38) terms a "cognized environment." The Bashu world is divided between opposing spheres of existence. On the one hand is the world of the homestead, in which the Bashu live, grow their crops, and keep domesticated animals. Surrounding this world, and impinging upon it, is the world of the bush, inhabited by the untamed and chaotic elements of nature, including powerful medicines and spirits. While these worlds are ideally separate, the continuity and productivity of the homestead depends on the performance of certain ritual actions that mediate between these two worlds and bring them into contact with one another on specific occasions. This mediation permits the domestication and incorporation into the homestead of certain spirits, medicines, and elements of nature, which are essential to the productivity of the homestead but are associated in their natural state with the chaotic world of the bush. Ritual mediation also serves to purge the homestead of untamed forces of the bush that have penetrated the homestead, causing misfortune, sickness, and famine. In other words, ritual mediation temporarily resolves major contradictions within the Bashu view of the natural environment in which they live and in which rain, spirits, and medicines are at once necessities of their existence and potential sources of misfortune.

While the performance of rituals designed to mediate between the homestead and bush is in the hands of local ritual leaders—clearers of the forest, healers of the land, former chiefs, and rainmakers—the activities of these leaders are ideally subordinated to those of the *mwami w'embita*, who, through his accession to office, acquires a permanent medial position between the homestead and the bush. This position, combined with his control of local ritual activity, permits him to insure the domestication of potentially beneficial forces of the bush and periodically to cleanse the chiefdom of chaotic forces of nature. Thus, the *mwami* is ultimately responsible for mediating between the opposing spheres of Bashu existence and for resolving the antinomies of Bashu experience. It follows that the death or weakness of the *mwami* results in misfortune and famine. The *mwami* thus resembles both the Dinka master of the fishing-spear, who brings together in a ritual context the divided worlds of man and divinity for the benefit of the community, insuring his followers of green pastures and sleek, fat cattle (Ray 1976: 117), and the Swazi king, who, through the annual *ncwala* ceremony, mediates between the world of the living and the world of supernatural beings, taking on to himself the "filth of the nation" and thus purifying and renewing his kingdom (Beidelman: 1966).

The central role of Bashu chiefs in controlling experience and in maintaining the vitality of the Bashu world makes the continuance of chiefship a vital interest for the ordinary members of Bashu society and explains the primacy of chiefs in Bashu social life. Moreover, this interest creates expectations concerning the ideal behavior of chiefs and candidates for chiefship and invests the actions of these political actors with cosmological importance, since any action that affects the chiefship may alter the balance between the homestead and bush, and thus the conditions of society. These expectations and assumptions about the nature of chiefly behavior, in turn, affect the ways in which chiefly actors operate in seeking to acquire and maintain political authority—specifically, their choice of allies, the definition and use of political resources, and the timing of political confrontations and encounters. They thus act, to use

Weber's image (1958: 280), like a railway switchman, who does not necessarily define the goal or destination one wishes to reach, but determines the tracks one follows in reaching this goal. So too, they define how chiefly behavior is interpreted and given meaning by society in general. In short, Bashu perceptions of the nature and purpose of political behavior play a major role in shaping Bashu politics. To understand politics among the Bashu, and Bashu political history, therefore, one must take account of how the Bashu themselves conceive of politics, what they, as opposed to the outside observer, believe they are doing in following particular courses of action. This, in turn, requires an examination of the symbolic function of chiefship as it is situated within the Bashu view of their natural and social environment.

The present study explores patterns of political competition among the Bashu during the nineteenth and early twentieth centuries by relating political process to the actor's own view of the interrelationship of environment, cosmology, and chiefship. Specifically, the study examines the competitive political processes involved in the establishment of the present line of Bashu chiefs in Isale and neighboring regions during the early years of the nineteenth century; the subsequent fragmentation of their political authority, leading to the formation of a series of related and yet politically autonomous chiefdoms; and finally, the impact of long-distance trade, the use of firearms, and the imposition of colonial rule on local patterns of competition and development at the end of the nineteenth and beginning of the twentieth century. The study argues that while these political processes occurred with considerable frequency among the peoples of eastern Africa and conform to patterns of political behavior that in fact transcend cultures, the specific form that they took among the Bashu was shaped by indigenous ideas about the nature and function of chiefship and about the relationship of chiefship to the changing conditions of the wider world.

The study thus employs a perspective toward sovereignty previously described by Beidelman (1966) in his pioneering study of Swazi royal ritual. By relating the symbolic structure of royal ritual to the purposes for which ritual is performed, Beidelman shows that the king is responsible for the reproduction of the social order. This onerous task is both the king's work and the basis of his pivotal position in Swazi society. The present study attempts to build on Beidelman's analysis by situating this prespective within a changing social and natural environment and by showing how the Bashu responded to political and natural crises in terms of both their political interests and their perceptions of the chief's "work."

By stressing the importance of cosmology, I do not mean to suggest that Bashu political concepts were an independent variable, removed from the social and economic forces that surrounded them. To the contrary, Bashu ideas about the nature and function of chiefship were themselves modified and refined by the forces and events that occurred during the nineteenth and early twentieth centuries. There was in a sense a dialectical relationship between ideology and action over time. The study therefore attempts to describe the evolution of Bashu political ideas as well as the impact of these ideas on political process.

While the Bashu provide the central focus of this study, they are by no means unique in their view of the nature of sovereignty. The insightful studies of Kuper (1947), Evans-Pritchard (1962), Beidelman (1966), Young (1966), Feierman (1972), and Mworoha (1977) on the ideology of African kingship have shown that kings were frequently defined by the members of society as ritual mediators between society and the forces of nature, and that, like Bashu chiefs, they were closely associated with the well-being of land and society and with the problem of ecological control.

It is equally apparent that these indigenous perceptions about the sacral nature of sovereignty affected the character of political action elsewhere in Africa. Thus, John Beattie (1972: 95), discussing kingship among the Banyoro of Uganda, states that "the ways in which people thought about the kingship, the *obukama*, had important implications for their behaviour, and the social relationships centering on it could only be adequately understood when account was taken of how the parties to these relationships thought about them and about themselves." Studies of political competition and change in a number of other African societies have illustrated this conclusion. Gerald Hartwig (1976: 144–49), for example, states that among the Kerebe of northwest Tanzania, kings were expected to regulate rainfall and that the inability to "conform to these expectations over an extended period of time was a major reason for deposing an omukama." Two kings are said to have been deposed in this manner at the beginning of the nineteenth century: Ruhinda, who was unable to prevent an excessive amount of rain from falling, and his successor, Ibanda, who fell victim to an extended period of drought.

Other studies have suggested that the belief in regicide found in a number of African kingdoms reflects the existence of a similar ideological relationship between political authority and the problem of ecological control. Thus, Evans-Pritchard (1962: 82–83) writes that among the Shilluk of the Nilotic Sudan, "The assertion that a sick or old king should be killed probably means that when some disaster falls upon the Shilluk nation the tensions inherent in its political structure become manifest in the attribution of the disaster to his failing powers. The unpopularity which the national misfortune brings on a king enables a prince to raise rebellion." In other words, while rebellions may reflect structural tensions within the Shilluk political system, the ideology of divine kingship in combination with a national disaster could determine the timing of encounters between political rivals and lead to a king's downfall.

Michael Young (1966: 151) reaches a similar conclusion concerning the practice of regicide among the Jukun of Nigeria:

> While the only documented case of a Jukun king's death in which regicide almost certainly occurred gives prominence to the political machinations of rival kinsmen, this king's death also coincided with a year of serious drought and famine conditions. Tradition and the mandate of the people's expectations underlay the political motives for the king's removal, and if it was for the conspirators a political coup, it was for the ordinary people of Wukari a validation and re-affirmation of their beliefs in the nature of kingship and the mystical association between the life of the king and the fertility of the crops.

Thus for the Jukun, as for the Shilluk and Kerebe, ideas about the ritual nature of kingship affected patterns of political competition.

Expectations concerning the ability of chiefs to control land also encouraged political leaders and competitors for authority to acquire resources and support that would bolster their ritual powers. Thus, Kerebe kings are said to have made a concerted effort to acquire special knowledge and medicines (Hartwig, 1976: 76). Similarly, among the Shambaa of northeast Tanzania, kings and competitors for kingship sought to monopolize the control of rainmagic over the land because, according to Steven Feierman (1974: 87–88), "Rainmagic is thought not to work unless one person has an effective monopoly over the magic for a given area." Thus, for example, Mbegha, the founding ancestor of the Kilindi dynasty that succeeded in establishing its authority over the Shambaa during the eighteenth century, made a specific point of confiscating the rainmagic of preexisting chiefs.

For similar reasons, competitors for political authority frequently sought the support of existing ritual leaders whose assistance was seen as essential for maintaining the well-being of the land, and thus, from an indigenous viewpoint, for the maintenance of political legitimacy. This pattern of support-building occurred in Alur (Southall, 1953), Bunyoro

(Berger, 1980), Rwanda (Vansina, 1962a), and among the Bemba (Roberts, 1973), Lunda (Schechter, 1976), Ndebele (Ranger, 1967), and Tallensi (Fortes, 1940), to name but a few examples.

The literature on African political systems in fact is filled with references that attest to the role of indigenous political values and of cosmology in general in shaping patterns of political action.[5] Yet as numerous as these references are, the historical relationship between cosmology and action has yet to be explored in a comprehensive fashion. This is largely because previous studies of political process and history have failed to examine in sufficient detail indigenous political ideas and the relationship of these ideas to the wider world view of the peoples being studied. In most cases the discussion of the cosmological dimension of sovereignty has been limited to observations concerning the general relationship between political authority and the problem of maintaining the fertility of the land, and has not paid adequate attention to either the complexity or pervasiveness of this linkage between politics and cosmology.[6] Failing to explore the depth and breadth of indigenous political perceptions has, I suggest, encouraged a fragmented view of the impact of these perceptions on political action. While indigenous ideas about the nature of sovereignty are shown to come into play on specific occasions —the most important of which involve periods of ecological crisis—at other times, cosmology appears to have little significance for politics, and both motivations and actions are ascribed either implicitly or explicitly to universal categories of thought and behavior as defined, for example, by Bailey (1969), Sahlins (1963) and Swartz et al. (1966). Political actors thus appear to take on and divest themselves of cosmological notions as the situation dictates, operating at one moment by universal rules of political behavior, and at others in a culturally defined mode. Yet, as Evans-Pritchard (1937) suggests in his study of Zande witchcraft, an individual's perceptions of the world cannot be easily laid aside, coming into play on certain occasions but not on others. Instead, these perceptions serve as a permanent (albeit changing) lens through which experience is interpreted and actions are given form. This being the case, indigenous political perceptions must be seen as coloring the whole political process and not just sporadic episodes. This is, in fact, the lesson of the Bashu experience.

By examining the relationship between cosmology and competition, the internal dimension of Bashu politics, the study hopes to demonstrate the necessity of understanding African politics and political history from the perspective of the actors involved. At the same time, it is hoped that the study will suggest new ways of understanding the nature of politics and political change in societies like the Bashu, in which an ideological equation existed between political leadership and the problem of ecological control. The list of such societies includes most of the precolonial chiefdoms and kingdoms of eastern and southern Africa.

RECONSTRUCTING THE POLITICS OF RITUAL CHIEFSHIP

Exploring the relationship between cosmology and politics during the nineteenth and early twentieth centuries required both a thorough examination of historical ideas about the nature and function of chiefship and a detailed study of patterns of political competition and change. In the absence of extensive written records, both lines of inquiry involved the use of methodologies and approaches that need to be defined at the outset, so that the reader is able to evaluate the arguments that follow. I have, however, reserved discussion of the methodological problems involved in reconstructing Bashu political ideas for chapter one, in which these ideas are described in detail. For it is difficult to understand the methodological arguments related to this problem without first having a grasp of the ideas being reconstructed. The present discussion will, therefore, center on the problem of studying historical patterns of political competition and development among the Bashu. I should note, however, that while this methodological division is analytically useful, the two lines of inquiry were closely related in actual practice. Thus, analyzing the ideology and structures of chiefship provided a methodological key for overcoming some of the difficulties inherent in the use of Bashu oral traditions.

The Bashu have a proverb that states, "When the story of Isale is told, children will die of hunger on the backs of their mothers." Isale is the center of the Bashu region, the area of oldest settlement, and an area that has experienced successive migrations, shifts in political leadership, intrigues, subversion, and succession struggles. The proverb suggests that to unravel all of this would take so long that the babies of mothers who attended the unraveling would die of hunger. It reflects, therefore, Bashu understanding of the complexity of their own history.

Bashu oral traditions, on the other hand, reveal a much more simplistic view of their history. They consist of short, informal, and largely unstructured descriptions of past events, including stereotypic tales of migration and settlement, references to local feuds and famine, stories about past chiefs, and descriptions of raids initiated by groups that occupied the upper Semliki Valley at the end of the nineteenth century. Each tradition is the property, and collective creation, of a single social group, the localized lineage, or, in the case of chiefs, guardians of the royal tombs, and is designed to represent the interests of that group or chief. The traditions are, therefore, myopic, in that they provide a parochial view of Bashu history, and reductive, in that they present only those elements of the past that inform contemporary social relations. Even the myth of Muhiyi, a genesis tradition describing the coming of the present line of Bashu chiefs, which is told by most informants, occurs in a variety of forms, each reflecting localized interests. Oral traditions, therefore, cannot by their very nature reveal the complexity of historical experience captured in Bashu proverbs.

As an historian, I was faced with the problem of reconstructing the complex historical experience reflected in the proverb of the "hungry babies" from the selective and myopic material presented in Bashu oral traditions. I began by collecting local traditions from a wide variety of groups, with the intention of forming a composite picture of their collective history. To this end, I conducted some 170 interviews and held numerous other informal discussions with Bashu men and women during the course of my field work.[7]

My first inverviews were with the present heads of the various Bashu chiefdoms and their court officials, including guardians of the royal tombs. These interviews provided an outline of the official court histories of the various chiefdoms, as well as certain insights into the nature of Bashu chiefship as perceived by the chiefs themselves. During the course of these interviews, I also collected the names of each chief's major subjects *(basoki bakulu)*. This list provided a second group of potential informants. The historical traditions presented by members of this second group closely resembled those I had been given at the royal centers. There were, however, some important exceptions, for a number of these subjects were members of commoner lineages. As such, they had their own traditions and view of the history of the chiefdom in which they lived. The testimonies of certain of these informants, moreover, conflicted with those of the courts in two ways. First, they challenged the frequent claims of Bashu chiefs that their ancestors had been the first occupants of the Bashu region, by claiming that their own ancestors had preceded the chiefs into the region. Second, certain commoner groups rejected the official claim that the ancestors of the present chiefs were the first chiefs among the Bashu. Instead, they stated that their own ancestors had been chiefs prior to the arrival and establishment of the present line of chiefs. To substantiate this claim, they provided the names of their major subjects, who, they said, continued to pay them tribute, a portion of which they now took to the present ruling chiefs.

Collecting the names of these subjects provided a third pool of potential informants. It also produced more contradictions. For while these new informants confirmed many of their patrons' claims, some of them claimed that their own ancestors had been the first occupants of the land and that they too had once had their own chiefs. Finally, as I moved out from this hierarchy of chiefs and subjects, interviewing men and women who were tied to various chiefs or former chiefs through marriage or the performance of certain political or ritual functions related to chiefship, the picture became even more confusing. I uncovered an ever increasing number of first occupants and former chiefs, to the point that it began to appear that there were as many first occupants among the Bashu as there are New Englanders who claim that their ancestors came over on the Mayflower, and as many chiefs as subjects.

Sorting out these various claims and piecing together a reliable picture of how these groups were actually related to one another historically seemed at first to be an insurmountable task. For it was difficult, if not impossible, to evaluate each group's claims from the traditions alone. It was here that the study of Bashu ideas about the nature and function of chiefship and the social relations of chiefship, collected simultaneously with Bashu historical texts, intersected with the study of Bashu political history to provide a key for resolving some of the contradictions imbedded in Bashu oral traditions. As noted above, Bashu chiefs were ideally viewed as ritual intermediaries between the world of the homestead and the forces of the bush. Yet this mediation required the cooperation of local ritual leaders, including the descendants of the first occupants of the land and former chiefs, who performed rituals that brought the world of the homestead into contact with the world of the bush. Without this cooperation, mediation could not be achieved and the well-being of the land would be disrupted.

This world view has insured the preservation and incorporation of earlier sources of ritual authority, going back to actual first occupants of the land, within the present structures of Bashu chiefship. Thus, for example, in performing sacrifices for the land a chief must be assisted by representatives of all the families who previously had influence over the land. Moreover, the actual order of settlement of these groups is reflected in the order in which they perform their ritual functions at this time. Similarly, a chief's investiture must be preceded by those of former chiefs and clan leaders who preceded the chiefs into the region. These prior investitures also reflect the order of each group's settlement. Finally, while all groups presently make an annual payment (*muhako*) to the chiefs in recognition of the chiefs' claims to first occupancy, the chiefs themselves pay a token *muhako* to actual first occupants of the land. Thus the social relations of chiefship and the performance of royal ritual provide a map for identifying and defining the actual historical relations that exist between various groups in the Bashu region, and for verifying or disproving claims made in the oral traditions of each group.

The use of extant ritual relationships as an historical tool can be seen in the following example of traditions that describe the settlement of the Bunyuka region of Isale. Since the history of this settlement forms the central episode in the myth of Muhiyi, which describes the founding of the Bashu chiefdoms, variants of the story are told throughout the region. The events described in each variant of the myth, however, are arranged in such a way as to support the interests of the group telling the story.[8]

The first two variants are from the descendants of Kavango, the ancestor of the present line of Bashu chiefs. These variants support the political position of the present chiefs and have a wide distribution.

> Muhiyi lived with his father, Kavango, in the village of Kavarola in Uganda. Kavango had many cattle which Muhiyi herded. One day Kavango noted that the milk which was placed in a special hut for him was missing. This continued to happen for several days. Finally, Kavango accused Muhiyi of having stolen the milk. Muhiyi, although innocent, was afraid of his father's anger. He therefore took his spear and hunting dogs and set out across the plains. Muhiyi followed the tracks of the buffalo and when he reached the Kalemba [Semliki] River crossed by means of a fallen tree. [In some versions of this variant he crosses a ford, in others he crosses on the back of a serpent.] Muhiyi continued his journey until he reached the foot of the Mitumba Mountains at a place called Kaviro. There, he succeeded in killing the animal he had been tracking. Muhiyi butchered the animal and dried its meat. He left some of the meat on a rock and the rest he carried home to his father. On arriving home he presented the meat to his father, who welcomed him for he had discovered that a serpent had drunk his milk. Muhiyi then told his father about the new land he had discovered on the other side of the Kalemba. His father said that he wished to accompany Muhiyi to this new land for there was no more room in their present land. Together they set off across the plains. When they reached the foot of the mountains they settled at Kaviro. Kavango then gave Muhiyi the land of Bunyuka in the mountains as a reward for having discovered the new land and for having given him the meat

from his kill. Kavango himself settled at Kivika. Together Kavango and Muhiyi chased away the Basumba who lived in the mountains and then cleared the mountains of forest.

Variant 2: Kyavambe lived in Kitara. He liked meat very much, but there were no animals left in Kitara. So he sent his servant Muhiyi to search for game. Muhiyi took his dogs and spear and crossed the Kalemba. On the other side of the Kalemba he found a beautiful country with many animals to hunt. He did not wish to return to Kitara and so took the route to Vutungwe (Bunyuka). Later Kyavambe took all of his family along with his cattle, sheep, and goats and crossed the Kalemba. He eventually arrived at Musagya Muguru, a hill in the plains. It was there that Kyavambe's wife gave birth to Kavango. Meanwhile, Muhiyi thought of his master and set out to return to Kitara. When he reached Musagya Muguru he met Kyavambe, who asked him why he had not returned. Muhiyi begged his pardon and said that he had found many animals. Muhiyi then killed two *mbara* for Kyavambe. After this Muhiyi returned to Bunyuka, which Kyavambe gave him in return for the *mbara.* Kyavambe died at Musagya Muguru and Kavango took the route to Kaviro and then Kivika, where he was invested *mwami w'embita* after his father's death.

The third variant comes from the descendants of Mutsawerya, who is locally identified with the 'hunter-king' Muhiyi. The distribution of this variant is limited to the chiefdom of Bunyuka, over which Mutsawerya's descendants presently rule while recognizing the paramountcy of Kavango's descendants.

Variant 3: Muhiyi lived in Kitara with his father, Kyavambe. Kyavambe owned many cattle, and Muhiyi was in charge of herding them. One day the milk which was placed in a special hut for Kyavambe began to disappear. Muhiyi's older brother, Kavango, accused Muhiyi of stealing the milk. Muhiyi, in his anger, took his dogs and spear and left his father's home. After he had left, it was discovered that the milk continued to disappear and that it was being drunk by a serpent. Kyavambe ordered a herdsboy to destroy the hut and the serpent. After this the milk no longer disappeared. Meanwhile Muhiyi had crossed the Semliki tracking a buffalo. When he reached the hill named Kaviro, he succeeded in killing the animal. Muhiyi saw that the land in the mountains was good and he decided to settle there. At that time there were Basumba living in the mountains, which were covered with forest. Muhiyi chased away the Basumba and began to clear the forest for planting. Muhiyi was later joined by Kavango and other groups of Bashu. Muhiyi gave them land and they gave him goats in return. Muhiyi took a woman from the Bito clan as his first wife and she bore him Kisoro, who succeeded Muhiyi.

The fourth major variant comes from the Bito (sing. Mwito) clan of Bunyuka, who claim to have welcomed Muhiyi when he arrived in Isale. The distribution of this variant is limited to the Bito of Bunyuka and to members of this clan living outside Bunyuka.

> *Variant 4:* [This variant follows the basic plot line of variant 3 up until Muhiyi arrives at the foot of the Mitumba Mountains. It then continues]: Muhiyi killed the animal he was tracking. He then built a fire and began roasting the meat from his kill. At this time there was a man named Sine, a Mwito, living in the mountains above where Muhiyi had camped. Sine saw the smoke rising from the plains below and took some of his men to go and see who was there. He found Muhiyi eating meat. Since he and his men were hungry, Sine asked Muhiyi for some meat. Muhiyi distributed the meat from his kill and they all ate together. After they had eaten, Sine noticed that Muhiyi was living in the open without a hut. He therefore invited Muhiyi to come and stay in his village in the mountains. Muhiyi agreed and accompanied Sine to Vungwe. Muhiyi stayed with Sine for a long time and Sine gave him some land on which to grow crops. He also gave him one of his daughters as a wife. This wife bore a son named Kisoro, who was invested *mwami w'embita* of Bunyuka.

A comparison of the myth's major variants reveals that each group arranges the narrative elements of the myth to support a claim to first occupancy, while undercutting similar claims made by other groups. Since the status of first occupant confers on its holder considerable ritual authority over subsequent occupants, the establishment of this claim is important. Accordingly, the Bito emphasize the role they played in welcoming Muhiyi and thus establish their prior occupancy vis-à-vis Muhiyi. Muhiyi's descendants, for their part, counter this Bito claim by populating the Isale with 'Basumba,' forest dwellers, whose claim to first occupancy is limited because they did not clear the land. They also assert their claim to prior occupany vis-à-vis Kavango's descendants by stating that Kavango followed Muhiyi and received land from him. Finally, Kavango's descendants attempt to establish their own claim to first occupancy by: (1) populating Isale with Basumba; (2) having Muhiyi return to Kavango before actually "settling" in Isale, so that Kavango and Muhiyi settle in Isale at the same time; and (3) having Kavango give land to Muhiyi.

In the second Kavango variant the process of undercutting the claim that Muhiyi preceded Kavango to Isale is more complete. Here, Muhiyi is a servant of Kavango's family who is sent to find a land with more animals. This reduces the independence of Muhiyi's actions and implies that if he discovered Isale, he did so on the order of Kavango's family. This may be a later variant reflecting an increase in the authority of Kavango's descendants over those of Muhiyi.

The actual order in which these groups settled in Bunyuka can be determined by checking the traditions of other groups less involved in Bunyuka politics and by examining the ritual relations that presently exist among these three groups. This combination of evidence indicates that the Bito were, in fact, the first occupants of the land. Not only do other commoner groups in the region refer to them as *bakonde,* literally 'clearers of the forest,' and thus first occupants, but the Bito perform a number of ritual functions that attest to the antiquity of their settlement. Thus, whenever Muhiyi's descendants wish to bury a member of their family or build a hut, they must first notify the Bito, who choose the location of the grave or hut and, in the case of a hut, place the central hut post. In addition, the Bito must perform the first sacrifice in Bunyuka at the time of the annual planting ceremonies. These ritual roles are traditionally assigned to the descendants of the first occupants of the land, who must be consulted before any action involving the land is performed.

In a similar fashion, it is clear that Muhiyi's family preceded that of Kavango, despite the contrary claims made by Kavango's descendants. To begin with, the central position of Muhiyi in all four variants suggests that the myth initially evolved out of the events surrounding the establishment of Muhiyi's family as chiefs in Bunyuka, and that it was later adopted and transformed by Kavango's descendants to legitimize their own claims to authority. The prior settlement of Muhiyi's family is also attested to by the following verse in a number of Bashu narrative songs, which are sung throughout Isale: "The day the chiefs left Kitara, they came one by one. Muhiyi was the first to leave."[9] More significantly in terms of the present discussion, Muhiyi's prior settlement is indicated by the ritual relationships that exist between his descendants and those of Kavango. First, Kavango's family must notify Muhiyi's descendants along with the descendants of other former chiefs before they can invest a new *mwami.* Second, Kavango's family give Muhiyi's descendants a goat every year. According to Muhiyi's (Mutsawerya's) descendants, this goat represents a token *muhako* payment in recognition of Kavango's having received land from Muhiyi. Kavango's descendants for their part acknowledge that the goat is given but deny that it represents *muhako.* However, the fact that they can give no other reason for why it is given, or for what else it signifies, undercuts their denial and reinforces the claims of Muhiyi's descendants.

Thus, understanding the ideology of chiefship and the social relationships that the ideology generates provided a means of defining the various stages of settlement and levels of authority in Bashu society, as well as the historical relations that existed between these stages and levels. It was therefore possible to resolve some of the contradictions presented by Bashu oral traditions.

Having defined the various levels of political and ritual authority in Bashu society, there remained the equally difficult problem of determining how these levels fit together over time. How and when were earlier sources of authority incorporated into the present Bashu chiefdoms? Here again the absence of written records posed a problem. Bashu traditions simplify the process of political domination, providing stereotypic explanations for how and why incoming groups established their authority over existing groups in the region. The traditions do, however, provide indirect evidence that can be used to reconstruct a picture of how and when various members of the present ruling group acquired the support of particular commoner groups and thereby expanded their political influence over a region. Particularly useful for this reconstruction was information on past marriage alliances and other historical relations, such as the roles one's ancestors played in the investitures and funerals of former chiefs.[10] While memory of such relationships declines as one goes backward in time, there was a remarkable body of data of this type going back to the beginning of the nineteenth century. By collecting information on these relations from both commoner and chiefly groups, I was able to reconstruct the alliance networks of successive chiefs going back to the founding ancestors of the present Bashu chiefdoms.

Placing these chiefly networks into their proper historical sequence made it possible to determine when a particular group was incorporated into a chiefdom and thus provided a picture of the growth of the chiefdom. This reconstruction also indicated shifts and losses of support over time, as certain groups that appeared in the list of allies in one generation disappeared from this list in the next generation, often to reappear among the allies of a rival chief.

Finally, by combining this composite picture with information on the status of groups that a chief chose as allies, derived from examining each ally's traditions and present position within the chiefship, it was possible to define the types of support each chief sought to acquire. This examination helped define the criteria employed in choosing allies and support, as well as how these criteria changed over time.

This type of historical reconstruction required a great deal of time. No single informant, no matter how knowledgeable, could provide information on all the alliances formed by a given chief or commoner. Thus information had to be collected from a number of different sources (often miles apart), collated, and cross-checked, in order to build up an overview of any given alliance network. Because of the time required to carry out this reconstruction, it was necessary to limit the geographical focus of my field work to a relatively small area. Thus, I spent most of my research time in the central Isale region, the history of which makes up the core of this study. By concentrating on Isale, I was able to construct a detailed picture of the development of Bashu political relations during the nineteenth and early twentieth centuries.

The choice of Isale was dictated in part by chance, for it was within Isale that I chose to settle and consequently established my most extensive and strongest social ties among the Bashu. In retrospect, however, the choice proved to be fortuitous, for Isale was historically the center of the Bashu political world. It was here that the Bashu first settled when they moved into the mountains. So too, it was in Isale that the present line of Bashu chiefs first established their authority. Moreover, the later expansion of these chiefs beyond Isale was generated by events that occurred in Isale. Finally, Isale appears to have had an historical coherence throughout much of the nineteenth century. Thus the definition of my field of inquiry did not impose an arbitrary or contemporary frame of reference onto the study.

On the other hand, Isale was by no means a self-contained region, cut off from the wider world that surrounded it. I therefore spent shorter periods of time in neighboring Bashu regions as well as among related Nande groups on both sides of the Semliki Valley. This research provided information on the general history of the wider region of which Isale is a part, as well as comparative data on the nature of political competition and development among related groups of people. The latter material, while incomplete, indicates that patterns of political action that occurred in Isale and that are described in this study also occurred, with some variation, among other Bashu and Nande groups. Verification of this conclusion, however, must await the completion of more intensive research in these neighboring regions.

ORGANIZATION

The study that follows is divided into three sections. Part One, "The Ideology of Bashu Chiefship," describes Bashu ideas about the nature and function of chiefship, the relationship of these ideas to the Bashu view of

the wider world in which they live, and how these ideas are expressed in
the royal rituals that surround the accession and death of a *mwami*. It also
discusses the antiquity of these ideas. This analysis provides the conceptual framework for understanding Bashu perceptions of politics and political change.

Part Two, "The Foundations of Bashu Chiefship," describes the early
development of Bashu society and examines how this development was
related to that of the wider Rift Valley region, of which the Bashu region
is a part. This section also discusses the origins of Bashu chiefship and the
conditions that led to the political expansion of the present chiefs' ancestors into the Bashu region. Finally, Part Three, "The Politics of Ritual
Chiefship in Isale," examines in detail the relationship between the ideas
described in chapter one and patterns of political competition in Isale
during the nineteenth and early twentieth centuries.

TABLE 1
BABITO DESCENDANTS OF KAVANGO

PART ONE

THE IDEOLOGY
OF BASHU CHIEFSHIP

FAMINE, PLENTY, AND THE LIFE CYCLE OF *BWAMI:* A BASHU VIEW OF CHIEFSHIP AND POLITICS

To understand the ways in which the ideology of ritual chiefship affected patterns of political competition and development during the nineteenth and early twentieth centuries, we need to define in concise terms how the Bashu themselves viewed chiefship during this period. For it was not simply the association of political authority with the problem of ecological control that determined the character of Bashu politics, but rather the particular way in which this association was perceived and defined by the Bashu.

In order to reconstruct Bashu concepts of chiefship as they existed in the past, it was necessary to begin by defining how the Bashu presently view chiefship. This was, of course, a difficult task, for the Bashu rarely, if ever, consider the ideas and values attached to chiefship as part of a unified system of thought. It was, however, possible to obtain a fairly clear image of how the Bashu view chiefship and of the relationship of chiefship to their wider world view by examining two types of data. The first consisted of direct statements about chiefship provided by Bashu elders and expressed in their historical traditions, proverbs, folktales, and —as noted in the introduction—in their explanations of natural phenomena. Also included in this category were the interpretive statements of informants concerning the meaning and purpose of Bashu royal rituals.

The second type of data consisted of indirect statements about chiefship, which were expressed through the performance of customary actions during Bashu royal rituals and specifically through the meanings of symbols in accession and funeral rites. These meanings are institutional. They inhere in the logic of customary forms of action and were discovered through a careful analysis of the specific context in which these symbols and actions occur within royal rituals, and of the relationship of these symbolic elements to the wider field of Bashu symbolic action, in other words, through an analysis of both royal and nonroyal ritual behavior.

In actual practice it was easier to examine nonroyal ritual behavior than to study the royal rituals of accession and death. This is because no chiefs were invested or buried during the period of my fieldwork. In fact, neither of these rites has been performed since 1950. I was, however, able to collect detailed descriptions of these royal rituals from a number of informants who had witnessed them in the past. Among these informants were three *basingya*, officials who had directed the accession or funeral of a *mwami*. These ritual leaders provided very rich descriptions of the ceremonies and preparations that accompanied these critical events in the life of a chiefdom. I have no doubt, however, that certain parts of these ritual processes were omitted. Other informants who had participated in one or both of these royal rituals in the past provided descriptions of the role they played and of the parts of the rituals they witnessed. These descriptions generally confirmed those of the *basingya* and in a few cases added details that the *basingya* had either omitted or hidden. The willingness, and even eagerness, with which informants talked about these rituals, often turning conversations about the activities of chiefs into discussions of these central rituals of chiefship, was striking and indicates, I believe, the importance of these rituals for Bashu chiefship, as well as the degree to which Bashu politics is concerned with ensuring the proper performance of these rituals.

While informant accounts could not totally replace firsthand observations, they did provide sufficient data to construct a detailed picture of the major ritual processes involved in the accession and death of a *mwami*. This picture, in turn, allowed me to examine the indirect statements about Bashu chiefship imbedded in these rituals.

Having constructed a picture of how the Bashu presently view chiefship, there remained the problem of determining the degree to which this picture was relevant to earlier periods of time or simply reflected contemporary ideas. Here again, I was faced with a difficult problem, for in the absence of written records, it is difficult to define historical points of view. There was, however, indirect evidence that attested to the antiquity of the ideas and practices obtained from contemporary sources. Since some of this evidence requires a prior understanding of the ideas and practices in question, it is best presented after they have been examined.

ECOLOGY, COSMOLOGY, AND CHIEFSHIP

Bashu chiefship, *bwami*, can only be understood when examined within the context of the physical environment within which the Bashu live, and in terms of Bashu perceptions of the forces that affect this environment. For chiefship is, above all else, a means of controlling these forces, and, by so doing, insuring the well-being of society.

The mountain ridges on which the Bashu cultivate their crops represent some of the richest farmland in eastern Africa. Blessed with rich soils and plentiful rainfall, averaging over 1300 mm a year, this land has traditionally given forth extensive crops of millet, beans, sorghum, and plantain bananas. Thus Frederick Lugard (1893: 178), who visited the lower slopes of the Mitumbas in 1891, noted that "All the country was densely inhabited and cultivated," while an early Belgian reconnaissance report described the region as "very fertile with extensive, well tended, and watered fields on all sides."[1] It was this productivity, in fact, which first attracted Europeans to the region. It also accounts, in large measure, for the present Bashu population of over 186,000 people, with an average population density of around 150 people per square kilometer.[2]

While the land and climate of the Mitumba Mountains provide the Bashu with a rich environment for growing their crops, it is an environment that is subject to climatic change and other manmade and natural disasters, all of which can damage or destroy a crop and quickly turn plenty into famine. Famine is in fact an ever-present reality that threatens their existence.

The most important variable in determining the success or failure of a harvest is rainfall. For, despite the high annual average, the quantity, periodicity, and quality of rain can vary from year to year. These variations may cause crop failure and bring on famine.

Variations in the quantity of rainfall can cause serious problems, as seen in the history of the two famines discussed in chapters five and seven. However, the total annual amount is often less important than the periodicity of the rains. If the rains do not start on time, crops may not germinate successfully or may germinate late and therefore not reach full maturity. Similarly, if the rains start on time but last too long, the crops may not ripen properly. The extended rains that I witnessed in the spring of 1974 are a prime example of how periodicity is often as important as quantity. The average total rainfall for March through June over a ten-year period from 1960 to 1970 was 458 mm. The total for the same period in 1974 was 524 mm, a difference of only 66 mm. However, when the combined ten-year average for the normally drier months of May and June is compared with the combined rainfall for those two months in 1974, one sees 170 mm for the ten-year average, as compared to 288 mm for 1974. There were thus 118 mm more than normal in these two months. Since beans require a period of dry weather before harvesting, the number of days with rain is also important. For the ten-year average there were 21.7 days of rain during these two months. For 1974 there were 34 days with rain during the same period. For the month of June alone the difference is even more striking: 9.9 days for the ten-year average, 19 days for 1974.[3]

The quality of the rainfall is also very important. What is needed are long soft rains that last all day, watering the earth and plants slowly without flooding them. While a thunderstorm may provide the same amount of rain, it can cause localized flooding, which can, in turn, cause erosion in an area where planting is done on the sides of hills. Thunderstorms are also undesirable because they are often accompanied by high winds and sometimes hail. Wind and hail can destroy a crop of millet in a matter of minutes. Thus, in their invocations for rain, the Bashu invariably call for *mbula nzolo*, 'a soft rain.'

A successful harvest thus depends on a combination of sufficient rains of the right type and at the right time. Too much or too little rain, or rain when the sun is needed, or storms can damage or destroy a crop and bring on famine. Variations in rainfall, moreover, are not the only causes of famine. The Bashu claim that in the past their crops were periodically destroyed by locusts that descended from the north.[4] In addition, warfare, both internal and from outside the Bashu chiefdoms, could result in the destruction of crops and bring on famine.

Thus, despite the apparent lushness of their environment, the Bashu live in a world in which plenty and famine can and do follow one another in quick succession, a fact that is adequately documented in Bashu historical traditions. For the Bashu, this alternation, so crucial to their existence, does not occur by chance. Rather, plenty and famine reflect the changing relationship between culture and nature or, in Bashu terms, between the world of the homestead *(ka)* and the forces of the bush *(kisoki)*. This relationship, in turn, is affected by the actions of men.

The world of the homestead is not simply the place of habitation. It is the whole domestic world: the huts, clearings, gardens, pastures, and markets, and, in certain circumstances, the whole chiefdom. It is a world of order and harmony, domestic animals and plants, ritual purity and benevolent spirits, a world that is regulated by communal values of reciprocity and cooperation and by the constant cyclical passage of time. Surrounding the homestead is the world of the bush *(kisoki)*, containing the elements and forces of nature: wild animals and plants, dangerous spirits, powerful medicines, and climatic forces. These elements are both essential and potentially destructive to the continued existence of the "homestead." In their natural condition they are chaotic, erratic, and violent, and thus a threat to the world of the homestead. Misfortune, famine, and death are, in fact, the result of forces of the bush penetrating the world of the homestead in their natural state.[5] For example, sorcerers *(baloyi)* cause sickness and death by introducing substances from the bush into the homestead, planting them in or near the victim's hut or in his garden. Similarly, women may cause misfortune by being possessed by certain malevolent spirits that inhabit the bush. These spirits are then allowed to enter the homestead, where they cause the death of children or the infertility of their host. Both consequences threaten the continued existence of the homestead. On a larger scale, famine may result from the onset of wild storms, hail, or violent rains from the bush, or from the intrusion of locusts or wild animals, such as elephants or buffaloes, which trample crops. Famine may also result from the pollution of the land through contact with uncontrolled elements of the "bush," such as the corpse of someone who has had leprosy or epilepsy, diseases that are identified with the bush. Such people are not normally buried in the homestead. Instead, their bodies are left in the bush.

While the forces of the bush are a source of danger for the homestead, they are also necessary to its existence. Crops require rain and dry weather. Certain illnesses can only be cured through the use of wild plants and herbs from the bush. Moreover, certain spirits of the bush are seen as potentially beneficial to the members of the homestead. For example, the spirit Muhima can protect members of the homestead from misfortune; Kalisya can aid in the breeding and care of livestock; and Mulemberi can help protect children. To acquire the benefit of these forces, however, they must be modified and controlled or, in Bashu terms, 'domesticated' *(-humbirya)*, before they are introduced into the homestead.[6] For if this is not done and they are introduced in their natural state, they will cause misfortune and death. Domesticated rains fall softly all day, nourishing the crops, while rain in its natural state comes in violent storms and destroys crops. Medicines of the bush that have been domesticated cure illnesses; undomesticated medicines cause sickness and death.

For the Bashu, therefore, the well-being of the homestead—the fertility of the land and productivity of people and animals—depends on the strict regulation of the relationship between the world of the homestead and the forces that inhabit the bush. These forces must either be excluded from the homestead or modified and domesticated. Regulating the relationship between the homestead and the bush is the responsibility of a number of ritual officials.

In everyday life, healers and herbalists, *bakumu,* mediate between the homestead and the bush, using their knowledge to cure individuals who have become ill as a result of the penetration of the homestead by forces of the bush. *Bakumu* operate by applying medicines gathered in the bush and domesticated through their influence, or by exorcising the undomesticated forces of the bush from the victim and reestablishing the division between the homestead and the bush. It is interesting to note in this regard that the process by which a person becomes a *mukumu* begins with a period of sickness and deliriousness, followed by his or her disappearance for several days, during which he or she wanders in the bush and, according to two *bakumu,* establishes a relationship with the forces of the bush.[7]

A second group of healers deals specifically with the illnesses of the land. 'Healers of the land,' *bakumu b'ekibugho,* function in the same way as healers of people, using substances from the bush that have been domesticated to improve the fertility of the land, or driving out undomesticated forces that cause pollution and infertility. While some healers of the land perform both functions, others specialize in one activity or the other.

A third group of ritualists who mediate between the homestead and bush are rainmakers. Rainmakers employ special rainstones, pieces of white and black quartz crystals, to attract rain to the world of the homestead and to control it so that it nourishes rather than destroys the crops.[8] Rainmakers are also able to drive away unwanted rains and storms and thus, like healers of the land, reestablish the division between the homestead and the bush. Again like healers of the land, some rainmakers perform both functions while others are specialized in the art of bringing wanted rains or in driving away destructive rains.

Priests for particular spirits of the bush can aid the productivity and security of the homestead by performing sacrifices that mediate the relationship between these spirits and the homestead. The most important of these priests are those for Muhima, mentioned above, Mulumbi, a serpent who aids in cultivation, and, more recently, Nyavingi, whose assistance is sought in a wide variety of matters, from agricultural production to success in trading and warfare.[9]

Finally, each of these ritual mediators, and the Bashu in general, are assisted in their efforts to regulate the relationship between the homestead and the forces of the bush by the ancestors. The Bashu conceive of their ancestors as benevolent elders. They are, in contrast to the spirits listed above, members of the homestead. Consequently, ritual actions at the time of death and subsequent communications with the deceased are designed to reestablish or maintain contact between the living and the dead, rather than to keep the ancestors away from the homestead, as occurs in a number of other African societies. In this capacity the ancestors help protect the homestead from the untamed forces of the bush and thus from misfortune and famine. This assistance is, in fact, viewed as a critical element in the control and domestication of these forces. It is for this reason that all ritual specialists seek the assistance of their ancestors in performing their mediating roles.[10] Of particular importance to the well-being of the land are the ancestors who first cleared the land of forest, for in doing so they established an important bond with the land. Their cooperation, obtained through the invocations of their descendants, is critical for the performance of any action involving the land.

While each of these ritual mediators helps regulate the relationship between the homestead and bush and resolve the antinomies of Bashu experience, they must work in cooperation with one another in order to be effective, for the domestication and control of forces of the bush requires communal action. Independent action, or actions that violate the communal values that order the homestead, 'break the homestead' *(-tul'eka)*, in the sense of exposing it to the outside world, and thus make it vulnerable to contamination by the untamed forces of the bush.[11]

Rainmakers who operate alone bring unwanted rains and violent storms. Only by working together can these mediators bring the long soft rains needed for growing crops. Similarly, healers who act alone bring misfortune and death and are viewed as sorcerers, *baloyi*. The distinction between a healer and a sorcerer, in fact, rests on the belief that healers work in a communal setting in which the patient is accompanied by relatives and friends and in which the assistance of ancestors is sought, while sorcerers work alone.

'Healers of the land' must also work in unison with both the community at large and with other healers. Thus, in cleansing a ridge, the 'healers of the land' who have influence over the ridge must coordinate their activities so that the cleansing proceeds in a downhill direction. Otherwise, the cleansing will be unsuccessful.

The homestead may also be broken by actions that violate communal norms of behavior. Thus sorcery, for example, is said to result from antisocial actions by the victim—the refusal to aid one's father or share one's wealth with one's brothers. Such actions open the homestead up to sorcery. Similarly, a woman's violation of social prohibitions that are designed to restrict her susceptibility to possession by forces of the bush can result in the introduction of malevolent spirits into the homestead and thus in misfortune. Finally, violating communal values may also cause misfortune to befall the homestead by alienating the ancestors, who turn their backs on the homestead and open it up to attack.

Regulating man's relationship with the forces of nature, therefore, requires the maintenance of communal values and the cooperation of ritual specialists. Unfortunately, neither condition is easily achieved. People act as individuals and in their own self-interest. Moreover, ritual specialists are not exempt from local politics and lineage relations and are thus frequently involved in disputes over land, unpaid bridewealth, blood payments, and other debts. These disputes may inhibit corporate ritual action and result in the conflicting use of rituals. For in disputes of this nature ritual leaders are said to employ their influence as a political resource, refusing to perform their service when it is needed, or, as is often the case with rainmakers, working alone to bring down unwanted rain or destructive storms on the land of their opponent. Because the regulation and control of the forces of the bush is difficult to achieve, the Bashu world is continually threatened by famine and misfortune.

The Bashu view chiefship, *bwami*, within the context of this wider view of man's relationship to nature. The chief, *mwami w'embita*, is the primary mediator between the world of the homestead and the world of the bush. Through the *mwami* the mediating roles of the rainmakers, healers of the land, priests of earth spirits, and ancestors are consolidated, and the forces of nature domesticated. The *mwami* is also ultimately responsible for separating the uncontrolled and dangerous forces of the bush from the world of the homestead. Finally, the *mwami*, through the use of judicial and political authority, helps regulate social relations among his subjects and prevent actions that violate communal values and 'break the homestead.' Thus political authority, like ritual authority, serves to control the relationship between culture and nature.

During the *mwami's* reign, all land rituals are coordinated by the *mwami* and occur in conjunction with the *vuhere vw'ovusyano* ceremony, which is performed annually before the planting of eleusine and marks the beginning of the agricultural year. During this ceremony, the *mwami* invokes the blessing of his ancestors and distributes the *mikene*, special ritual seeds thought to insure the fertility of the land when placed among the gardens of the *mwami's* subjects. The *mwami* distributes these seeds to the heads of localized lineages whose authority he has recognized and to members of the royal family who serve as subchiefs over various areas of the chiefdom. These men return to their homes and coordinate the performance of local land rituals, which can only occur with the *mwami's* consent and blessing. The *mwami* also performs periodic ceremonies to cleanse the land of pollution caused by uncontrolled forces of the bush. These include the new moon ceremonies, rituals performed at the time of planting, and, most importantly, the *mwami's* investiture (see below). Finally, the *mwami* helps prevent actions that would break the homestead through the exercise of his political authority and through the levying of fines for violations of certain ritual prohibitions. Today this authority is supported by the force of the national government. Traditionally, however, the *mwami's* authority depended on his ability to withhold his ritual services and bring on misfortune or on his success in calling upon loyal supporters to pressure deviant individuals and groups into respecting his wishes.

In principle, therefore, there is no conflicting use of ritual during the *mwami's* reign. The ritual condition of the land is protected. Actions that might break the homestead are restricted, and famine is prevented. The power of chiefship, however, is not constant. Individual chiefs may die or be unsuccessful in coordinating and controlling local ritual leaders. When this happens, the use of ritual devolves to local officials, pollution accumulates, the homestead is broken, and there is famine. The power of chiefship, *bwami,* is thus seen as fluctuating through time. It is, in fact, viewed as an almost organic substance that grows, declines, and is reborn. As it does so, it affects the condition of the land and everything in it. Thus the waxing and waning of the land, cycles of plenty and famine, are directly tied to the waxing and waning of chiefship.

This relationship is reflected in Bashu concepts of time.[12] Bashu time, like that of other African societies, differs from Western time. It is multidimensional, with various rates and patterns of time associated with different social and economic activities, astrological movements, meteorological changes, and ritual processes. Time is episodic and discontinuous rather than linear, often moving in repetitive cycles that may be stopped or even reversed.

In general, Bashu time may be divided into two categories, what Evans-Pritchard (1940: 94) defines as 'ecological time' and 'structural time.' Ecological time is time determined by ecological relationships. It is a reflection of Bashu relations to their environment and is associated with economic activities and especially with agriculture. Structural time, on the other hand, is a reflection of social relations and is associated with the developmental cycle of domestic groups and ritual offices and with the passage of individuals through the various stages of life.

Within the context of Bashu chiefship, these two categories of time are connected. For the developmental cycle of chiefship, a passage through structural time has a direct impact on the passage of ecological time. The growth, decline, and renewal of chiefship provides the central temporal movement that guides and orders ecological time.

Ecological time is cyclical and is experienced in relationship to the passage of the agricultural year. It is thus measured by the ordered performance of activities associated with agriculture: clearing the land, planting crops, weeding and harvesting, and by the dry-season activities that occur between the fall and spring growing seasons and again between the spring and fall seasons. Because ecological time is tied to the agricultural year, its flow is determined by the regular movement of wet and dry seasons. This movement, in turn, is directed by the actions of men and specifically by the regular and coordinated performance of rituals designed to regulate the relationship between nature and society. In other words, the flow of ecological time is determined by the proper performance of ritual.

Ultimately, ritual activity is linked to the developmental cycle of chiefship. Thus, ecological time, the orderly passage of the seasons and the performance of agricultural activities, is subordinated to the temporal movements of chiefship. When chiefly power is strong, ritual control is maintained, and ecological time unfolds in regular cyclical movements, there is plenty in the land. Conversely, when chiefly power declines as a result of the death or weakness of a *mwami,* the coordination of ritual is broken, the movement of the seasons is disrupted, and agricultural activities are ceased. The orderly passage of ecological time gives way to the erratic temporal movements of the bush. Moreover, since all chiefs eventually die, the Bashu view of ecological time can be said to be entropic, in that time winds down with the decline of a chief's power, only to be revived and renewed with the accession of a new chief. From this perspective the accession ceremonies can be seen to revitalize society by renewing the flow of ecological time. Bashu concepts of time, therefore, reflect the central importance of the *mwami w'embita* for the well-being of the Bashu world.

Bashu ideas concerning the nature and function of chiefship are embodied in the symbols and actions surrounding the accession and death of a *mwami.* While I do not intend to provide a complete analysis of these rituals—a task that, in any case, is extremely difficult, given their esoteric nature—it is important to examine the major themes expressed in them to understand how the Bashu view chiefship and how this perception affects political action.

ACCESSION AND THE GROWTH
OF BWAMI

The life cycle of *bwami* begins with the accession of the *mwami w'embita*. However, since *bwami* is eternal, there can be no beginning or end, and the accession rites mark a transition between the interregnum, associated with disorder and famine, and the reign of the *mwami*, associated with order and plenty. Accordingly, it is a period of rebirth, growth, and maturation, ending in the renewal of *bwami* and of the land over which the *mwami* rules.

The actual accession of the *mwami* is preceded by a series of preliminary ceremonies designed to consolidate all ritual power over the land and thus to strengthen and nourish *bwami (-kul'obwami)*. These proceedings also emphasize the interdependence that exists between the *mwami* on one hand and ritually powerful commoners on the other. While it is the *mwami* who insures the productivity and security of the land, he can only do so with the cooperation of these local ritual leaders. In this sense, the power of *bwami* can be seen as a collective entity involving all ritual power over the land.[13]

The preliminary ritual activities begin with a visit by the *mwami*-to-be, or his representative, to the descendants of former chiefly families that once ruled the area over which the *mwami* is about to acquire control. Each of these families is asked to provide the *mwami* with a gourd of milk. By granting this request they acknowledge the *mwami*'s authority, pledge their cooperation and support, and symbolically nourish *bwami*. It is said that if this request is denied, the *mwami* cannot be invested.[14]

Once the *mwami* has received the recognition of former chiefly families, the second phase of the preliminary activities begins. This involves the investiture or reinvestiture of the heads of these families. These rites are themselves preceded by the investitures of still earlier families of chiefs and of various ritual figures throughout the chiefdom. Coordinating the investitures of these local leaders is viewed as important because each office has its own developmental or life cycle and, like *bwami*, waxes and wanes with the life of its occupant. The developmental cycles of these offices, however, do not necessarily parallel that of *bwami*. Thus, at the time of the *mwami*'s investiture, other offices may be at different stages of development: one may be held by a man whose authority is waning, another by a man who has not yet obtained full authority, and yet a third may be without an occupant. The ritual power of each office may therefore be at a different level of effectiveness. Investing, or reinvesting, the occupants of all offices together with the *mwami* brings the developmental cycles of each office to full strength at the time of the *mwami*'s investiture. By simultaneously strengthening and consolidating all ritual control over the land, the flow of ecological time disrupted by the death of the previous *mwami* is renewed and the chiefdom revitalized.

The temporal control of chiefship over the developmental cycles of subordinate offices also serves to define political relations among the Bashu. A chief or official whose office is temporally subordinate to that of another chief is politically inferior to that chief. Consequently, the process of political domination involves, from a Bashu viewpoint, the extension of the temporal dimension of chiefship over an ever-increasing number of other ritual/political offices.[15]

While preliminary ceremonies emphasize the coordination and consolidation of ritual power and serve to nourish *bwami*, it is through the actual accession of the *mwami* himself that *bwami* is fully matured and the chiefdom renewed. The symbols and ritual actions that surround the *mwami*'s accession, therefore, reflect themes of renewal and maturation.

The accession ceremonies begin with the appearance of the moon's first quarter, a time associated in Bashu thought with purification and rebirth. This association is seen elsewhere, in the identity that exists between the moon's first quarter and a woman's menstrual period. When the moon's first quarter appears, the Bashu say *mughenda amatwa*, 'the moon appears.' This same phrase is used as a euphemism for a woman's menstrual period, which the Bashu associate with cleansing, believing that it washes away the pollutive substances that have built up within a woman's body between menstrual periods. The association of the new moon with purification is also seen in the performance of rituals designed to cleanse a person who has been contaminated by forces of the bush during this same lunar phase. Finally, the performance of annual ceremonies for the purification of the land, as well as periodic ceremonies to renew chiefship, occurs during this lunar period.

As the *mwami* undergoes various rituals associated with the investiture process, the moon waxes, becoming completely full with the climax of the accession ceremonies and the presentation of the *mwami* to his people. The full moon, in general, is associated with maturation and wholeness. When it appears, the people say *mughenda amaghoma*, meaning that the moon has become round, in the sense of whole and complete. Like the moon, *bwami* has reached maturity. Thus when the newly invested *mwami* is presented to the people, one of the phrases with which he is greeted is *bwami bw'amaghoma*, '*bwami* is whole and complete.'

The maturation of *bwami* is further symbolized by the procedure used in preparing the beer that is drunk at the end of the investiture ceremony. When the new moon appears and the *mwami* begins his initiation, the officials who direct the investiture proceedings, the *basingya*, cut the bananas to be used in preparing the beer. In contrast to the normal procedure of cutting ripe bananas, these bananas have not yet reached maturity and are placed in a hut to ripen during the period in which the *mwami* himself is secluded in a special hut and receives his power. Their ripening therefore parallels the maturation of *bwami*. When the *mwami* reappears and receives his emblems of authority, the bananas are squeezed and fermented for three days. The fermented beer is then drunk at the feast celebrating the *mwami*'s presentation to his subjects.

The theme of purification, as well as the connection that exists between the condition of *bwami* and that of the land, is also reflected in the performance of the *vutambe vunene*, a ritual cleansing of the chiefdom, which occurs in conjunction with the *mwami*'s investiture. The *vutambe vunene* is begun with the appearance of the moon's first quarter. 'Healers of the land' *(bakumu b'ekihugho)* pass through each village in the chiefdom, driving out disruptive spirits that have occupied the land during the interregnum and in general cleansing the land of pollutant substances. The *bakumu* employ a bouquet of special plants, which is brushed along the pathways and over the doorways and gardens of each village. The bouquet is later taken to a valley and left in the bush along with a chicken, which is sacrificed to the *balimu vavi* to keep them from returning to the villages. In this way, the whole land is renewed along with *bwami*.

The conjunction of the investiture of the *mwami* with the lunar cycle, the brewing of beer, and the cleansing of the land helps to realign the flow of ecological time with the temporal movement of chiefship. It also emphasizes the processes of renewal and maturation, which are the central concerns of the accession ceremonies, as well as the relationship of *bwami* to the land. These themes are also reflected in the actual initiation of the *mwami*, a rite of passage involving the removal of his former status, a period of transition, and the taking on of the power and status of *bwami*.

The *mwami*'s initiation begins at sunset, following the appearance of the moon's first quarter, with his separation from the homestead and entrance into the world of the bush, where the initiation occurs. The timing and direction of this movement indicate that the *mwami* is undergoing a social death through which he sheds his former status. In Bashu marriage rituals, the separation of the bride from her natal homestead and her transfer to the homestead of her future husband, where she will begin a new life, occurs at sunset and involves a journey in which she passes through a section of the bush. Similarly, candidates for circumcision, an event that marks the transition to manhood, begin their rite of passage by leaving their homestead and entering the bush at sunset.

In preparation for the *mwami*'s entrance into the bush, the *basingya* cut a small clearing in which a hole *(vighala)* is dug in the shape of a grave. A cot is then placed in the grave. The *mwami* passes the night on this cot along with a ritual wife known as the *musumbakali*. The cot is fabricated out of the same plants that are used to cleanse the land during the *vutambe vunene*, which coincides with the *mwami*'s initiation. While in the context of the village, and thus the world of the homestead, these plants are associated with cleansing and purification, in the context of the world of the bush, where they are thrown and where the *mwami* receives his investiture, they are associated with pollution and danger, and specifically with the pollution of the chiefdom. By spending the night on a cot that has been fabricated from these plants, the *mwami* becomes extremely polluted and in effect takes on what Beidelman (1966: 396) has called, in the context of Swazi royal ritual, the "filth of the nation." By doing so, he completes the cleansing of the land, which began with the *vutambe vunene*, and insures the renewal of the chiefdom.

Spending the night in the *vighala* completes the *mwami*'s social death. In the remainder of the initiation process, he is cleansed of the dangerous forces he has come into contact with in the *vighala* and takes on the new status as the living embodiment of *bwami*.

The cleansing process in fact begins while the *mwami* is still in the *vighala*, with the partial sexual joining of the *mwami* and the *musumbakali*, in which the *mwami* withdraws his penis at the point of orgasm and ejaculates onto a green banana leaf. This procedure reflects Bashu ideas about pollution and sex. The Bashu maintain that a man who has become polluted through the performance of a ritually dangerous action, or through contact with a pollutant substance, may free himself of this pollution by having sexual relations with a woman and thereby transferring the pollution to her. Accordingly, a man who has committed incest, the epitome of antisocial action, must roam around in the bush, neither washing nor cutting his hair, living in fact like an animal, until he catches a woman and rapes her. A woman in similar circumstances must seduce a man. In cases of extreme pollution, however, the sexual act may result in the death of the unpolluted partner. In such cases a surrogate must be used. For example, the *mughula*, the official responsible for preparing a deceased *mwami*'s body for burial and for removing his lower jawbone (see below), is required to visit a healer before returning home, for he has by his actions become highly polluted and must be cleansed. The *mughula* passes the night out of doors at the healer's home, where he sleeps on green banana leaves. During the night he must ejaculate onto the leaves, which, according to one informant, are seen as a surrogate woman. Green banana leaves in general are used to 'cool' ritually dangerous substances, which are 'hot.'[16] The *mughula* can then return home, where he is required to have sexual relations with a woman he takes 'by force.' This woman then becomes polluted and must visit a healer to be cleansed. When I asked why the *mughula* must spend the night at the healer's before returning home, I was told that the burier's pollution is too powerful and would kill the woman with whom he first had sexual relations. It is apparently for this reason that the *mwami*'s purification begins with this act of coitus interruptus with the *musumbakali*.[17]

Early the next morning, when a cock placed by the *vighala* crows, the *mwami* and *musumbakali* leave the *vighala*. As the *mwami* ascends from the ritual grave, an official known as the *mwamihesi*, or *muhesi w'omwami*, 'the blacksmith to the *mwami*,' holds a hammer used for forging iron in each hand and beats them together. The *basingya* respond by asking, "Whom have we forged?" The *mwamihesi* strikes the hammers a second time and the *basingya* say the name of the new *mwami*. The *mwami* is thus compared to a piece of forged iron. Like the piece of iron, he has undergone a transition from one state to another. Moreover, like the piece of iron, the *mwami* is still 'hot,' still ritually dangerous.

The *mwami*'s head is then shaved. This marks the end of the period of mourning for the previous *mwami*, and, as in commoner funerals, contributes to his cleansing and reincorporation into the homestead. Shaving the *mwami*'s head, therefore, continues the process of purification begun in the *vighala*. It should be noted, however, that the *mwami*'s head is not completely shaven. A tuft of hair at the back of the head, called the *lisunza*, remains long and is never cut. The significance of this practice will be discussed below.

The *mwami*'s arms, chest, neck, and back are next covered with a mixture of resin oil and a red powder acquired from the forest to the west of the Bashu region. The *mwami* also receives a new set of clothes made of barkcloth that is covered with this same mixture. In other contexts the mixture of red powder and oil is applied as part of the cleansing process by healers to individuals who have been possessed by malevolent spirits or who have committed a ritually dangerous act *(-lolo)*. The color red is thus associated with cleansing and the application of the oil and powder contributes to the *mwami*'s purification. Red also signifies maturation and growth. The Kinande word for 'red,' *ngula*, is related to the word 'to grow,' *-kula*.[18] Thus the application of red powder also symbolizes the growth of *bwami*. Finally, the color red signifies liminality and danger. The *mwami* is no longer in the world of the *vighala* associated with extreme pollution and danger. However, he has not yet been cleansed and reincorporated into the world of the living. He is thus 'betwixt and between' (Turner, 1967), a condition to which he will periodically return in order to fulfill his function as the primary mediator between the world of the homestead and forces of the bush. At such times, the *mwami* will be reanointed with this same mixture of powder and oil.

The *mwami* remains in this liminal condition for seven days and cannot bathe or remove the red clothing. He is also kept in seclusion. The period of seclusion completes the cleansing or 'cooling' process and brings *bwami* to full power.

The period of seclusion is spent in a specially constructed hut, the sides of which are fabricated out of green banana leaves. The hut is similar to that in which newly circumcised boys stay following their circumcision. In both cases, the green banana leaves are thought to 'cool' the hut's occupants, who are in a state of ritual danger and thus 'hot.' During his seclusion in this hut, the *mwami*'s food is served in an unfired pot, which is 'cool' and contributes to the cooling process. In a similar fashion, newly circumcised boys may eat only cold food.

From the time he leaves the ritual grave to the time he ends his period of seclusion, the *mwami* is referred to as a little baby. When he leaves the grave the *basingya* say *twavirevuta olumekeke*, 'we have given birth to a little baby.' This is a complex image. On the one hand, newborn babies are thought to be ritually pure, or 'white,' in the sense of innocence, as opposed to adults, who have been blackened *(-yira)* through experience. Yet the use of the phrase during this period of seclusion and liminality evokes associations with the liminal condition of a newborn child prior to the falling of his or her umbilical cord. During this period, the child is between the world from which he came and the world into which he is born. Like the *mwami*, he must be secluded during this period, for he is in a state of ritual danger and is said to be highly susceptible to pollution. This association is made explicit by the common use of the term *bwami* to refer to a newborn child in this condition.

The ritual danger experienced by the *mwami* during his period of seclusion is transferred to the land and everything in it as a result of the *mwami*'s association with the physical and moral condition of his chiefdom. Consequently, a number of ritual prohibitions apply to the people living within the chiefdom during this period. These prohibitions are designed to protect the chiefdom from pollution.

To begin with, all sexual activity is prohibited, not only among people but among animals as well. It is said that if a person has sexual relations during this period, he threatens the life of the *mwami* and thus the well-being of the land. More specifically, it is felt that sexual activity will disrupt or negate the investiture process and by doing so prevent the renewal of *bwami* and the land.[19]

The use of iron tools, such as axes, hoes, and billhooks, is also prohibited during the period of the *mwami*'s seclusion. This is not a general prohibition against work, for tasks that do not require the use of iron tools may be carried out. It is even possible to till the land if a wooden stick is used. The prohibition thus applies to contact between the land, including trees and plants, and iron. If iron is brought into contact with the land during this period, famine will occur.[20]

A third prohibition applies to nursing children. The susceptibility of children to pollution decreases with age and "experience." However, until a child is weaned he is still viewed as vulnerable, and especially so during periods of ritual danger. During the investiture of the *mwami* all nursing children are required to leave the chiefdom or face the danger of sickness and death.

The *mwami* receives special medicines, knowledge, and sources of power associated with chiefship during his seclusion. These are provided by the head of the *basingya*. The actual process by which the *mwami* acquires these elements of chiefship is the most esoteric aspect of the accession rites. While I was able to collect scattered references to these proceedings, it is impossible to reconstruct them in any meaningful way. It is clear, however, that the *mwami* receives certain powerful objects, including the pieces of the jawbones of preceding *bami*, during this phase of the accession rites. Since the Bashu maintain that the power of *bwami* resides within the jawbone of an invested *mwami*, these relics represent the accumulated power of *bwami*. They are placed in a special sack, into which the *mwami* will later place the ritual seeds that he distributes during the annual planting ceremonies.

With the termination of the period of seclusion, the *mwami* is conducted from his hut to a special log, *mukoko*, on which he sits with his *mombo* (a ritual wife described below) and the *semwami*, who is also called the *mukulu* or *mubito* and represents the *mwami*'s eldest brother and head of the royal lineage. The *mwami* then receives his emblems and instruments of authority. The most important of these is the *mbita*, a crown fabricated out of the skin of a flying squirrel, which is covered with a cap, *kakira*, made of knotted raffia cord containing a variety of bones, teeth, and medicines.

The *mbita* is the most powerful symbol of Bashu chiefship. It is also a source of extreme pollution. The flying squirrel, *mbake*, is one of the several animals that fall between Bashu taxonomical categories, and it is classified as an *nyama ihalire*. The adjective *ihalire* is derived from the verb *-hala*, meaning 'to become soiled.' An *nyama ihalire* is thus an animal that causes or is a source of pollution. The teeth and bones used to make the *kakira* come from wild and savage animals, none from domestic animals. The leaves and twigs that are contained in the *kakira* come from a variety of wild plants, which, following the list of plants provided by an informant whose family is responsible for fabricating the *mbita*, includes several of the plants used in making the bed on which the *mwami* spends the night in the *vighala* and in cleansing the land. While the complete meaning of this mélange of materials requires an analysis of the meanings and uses of each object used to fabricate the *mbita*, a task that is beyond the scope of this study, there can be little doubt that the *mbita* is associated with the world of the bush and the forces that inhabit the bush and is seen as a great source of pollution. Thus, any contact between the *mbita* and the land will cause famine.

The pollution and danger associated with the *mbita* raises the question of why such an object is placed on the head of the *mwami*, who has for several days been cleansed and cooled of the pollution he acquired in the *vighala*. The wearing of the *mbita* would appear to be contradictory and counterproductive to this process. In short, why go to all the trouble of cleansing the *mwami* if you are going to contaminate him again by placing this mélange of ritual filth on his head?

Part of the answer to this question lies in the fact that the *mbita* is never allowed to come into direct contact with the *mwami*. The *mwami* may not touch it with his hands, and before it is placed on his head, he receives a cap made from eagle feathers, which are said to protect the *mwami*. Second, the *mwami* does not keep the *mbita* on his head for very long. It is removed shortly after this investiture ceremony and is only replaced on his head on special ritual occasions associated with the renewal of chiefship and the land.

While these restrictions limit the possible danger involved in placing the *mbita* on the *mwami*'s head, they do not explain why the *mbita* is made of such polluting substances in the first place. There appear to be several possible reasons for this. First, by wearing the *mbita*, a distillation of the forces of the bush, the *mwami*, who is otherwise protected from pollution, creates the condition of liminality that is essential to his role as mediator between the homestead and the bush. This interpretation is supported by the fact that red powder, associated with liminality, is applied to the *mwami*'s body whenever he puts on the *mbita*.

Second, the suggestion that some of the plants used to fabricate the *mbita* were used in making the cot on which the *mwami* spent the night in the *vighala*, and in the bouquet used in cleansing the land, implies that the *mbita* may also symbolize the "filth of the nation," which the *mwami* must take on each time the land is renewed.

Finally, the *mbita* is a sign of great power, evoking awe from all those who view the *mwami* wearing it. For the Bashu say that the destructive force of the *mbita* would kill an ordinary man even if he had the benefit of the eagle-feather cap. Only a true *mwami*, who has been properly invested and who has received the knowledge and medicines of chiefship, can safely wear the *mbita*.

The presentation of the *mbita*, along with a number of copper and iron bracelets *(miringa)* completes the *mwami*'s accession to power. His body is washed and he receives new clothes, including the skins of a leopard, an antelope called *ngabe*, and a civet cat. The *mwami* is prohibited from removing all of the red powder and oil placed on his body during the investiture procedures, however, and must retain some on his back and chest. This practice is evidently related to the above-mentioned practice of leaving a tuft of the *mwami*'s hair uncut throughout his reign. Together they indicate that the *mwami*'s condition remains somewhat liminal, that there is an inherent contradiction in his character. This contradiction is reinforced during the critical periods of renewal noted above but is never eliminated. It is perhaps because of this permanent liminality that the *mwami* must be protected from dangerous actions and substances during his reign by a special guardian known as the *muhangami*.[21]

The *mwami*'s investiture may thus be seen as establishing a permanent association between the *mwami* and forces of the bush. While this association is normally restricted and controlled by the prohibitions that shape his daily life, they can never be totally eliminated and are, in fact, strengthened from time to time so that the *mwami* may protect the land by mediating between the homestead and the bush.

The accession ceremonies conclude with the presentation of the *mwami* to his subjects, an occasion that is accompanied by great feasting, oaths of allegiance, the rekindling of the royal fires, and the performance of a sacrifice in which the *mwami* invokes the blessings of his ancestors and calls for *bandu bangi, syombene ningi, na kalyo kangi,* 'many people, many goats, and much food.' The *mwami* then receives a spear, a sacrificial knife, and a billhook. These are subsequently given to the *mukulu,* who serves as a ritual surrogate for the *mwami,* performing ritual acts, including the sacrifice of animals, which are ritually dangerous for the *mwami* in his liminal condition. The presentation of the *mwami* corresponds to the appearance of the full moon and marks the full maturation of *bwami.*

DEATH AND THE BREAKING OF BWAMI

The death of a *mwami* does not destroy the power of chiefship, for the Bashu maintain that chiefship is eternal, *bwami sivivitala,* '*bwami* never dies.' Instead, the *mwami's* death marks the end of one phase in the life cycle of *bwami* and the beginning of the transition to a new phase, the interregnum. This transition is effected by the performance of the mortuary rites. These rites therefore resemble the accession rites, in that they transform *bwami* and represent a liminal phase in its life cycle. There is, in fact, a close parallel between the actions and symbols of the mortuary rites and those of the accession rites. However, the two sets of rites are mirror images of one another and move *bwami* in opposite directions. The accession rites consolidate ritual power, strengthen and personify *bwami,* reinforce control over forces of the bush, and renew the land. The mortuary rites, conversely, break up ritual power, weaken and separate *bwami* from the deceased *mwami,* and open up the land to the uncontrolled forces of the bush. In short, the mortuary rites undo what the accession rites have done.

At the death of the *mwami,* those in attendance state *bwami bw'amatwika,* '*bwami* is broken.' The same term is used to describe the last quarter of the moon, *mughoma w'amatwika.* Thus the logic of lunar imagery encountered in the accession rites occurs again in the mortuary rites: *bwami,* like the moon, is waning. Royal fires are extinguished and the *mwami's* emblems of authority are removed. His body is then taken at night to a special hut, where it is kept in seclusion until burial. This part of the mortuary rites reverses the 'coming out' that ended the *mwami's* seclusion during the accession rites and that occurred during the day. It also reverses the *mwami's* receipt of emblems of authority and the lighting of the royal fires that followed the *mwami's* 'coming out.'

During the period of seclusion the *mwami's* body is placed on a cot, under and around which are placed smoking fires, which speed the process of decomposition. After seven days the *mughula* removes the *mwami's* lower jawbone. Again, this phase of the mortuary rites reverses the transformation that occurred during the *mwami's* previous seclusion. While the *mwami* was 'cooled' and purified during the accession rites, his body is heated, becoming a great source of pollution that cannot come into direct contact with the ground, during the mortuary rites. Moreover, while the *mwami* takes on the power of *bwami* during his initial seclusion, culminating in his receipt of pieces of the jawbones of former *bami*, this process is reversed during the mortuary seclusion with the removal of the *mwami's* own jawbone. The parallel between the two periods of seclusions is completed by the enforcement at this time of the same set of prohibitions that affected people living in the land of the *mwami* during his earlier seclusion.

Finally, the *mwami's* body is placed on a cot similar to that on which he spent the night in the *vighala*. This in turn is placed in a hollowed-out tree trunk,[22] which is then lowered into a grave, around which is built a fence of reeds that eventually takes root and surrounds the grave with bush. The reversal of the accession rites is thus completed with the return of the *mwami* to the grave from which he exited at his investiture, and from which, symbolically, his successor will also exit. A special drum beaten to signal the *mwami's* exit from the *vighala* now marks his burial and initiates the interregnum. By reversing the accession ceremonies, the mortuary rites express both the break-up of *bwami* brought on by the death of the *mwami* and the continuity of chiefship. At the same time, the mortuary rites preserve *bwami* by separating the office from the now moribund and polluted body of the incumbent. Finally, the death and burial of the deceased *mwami* reverses the temporal movement of chiefship. This in turn reverses the flow of ecological time and gives rise to the chaotic conditions associated with the interregnum.

The interregnum period marks the lowest point in the life cycle of *bwami*. Ritual control over the land is decentralized, the flow of ecological time is disrupted, and the land is 'uncovered,' opened to the uncontrolled forces of the bush that threaten society.[23] The period is consequently associated symbolically with the conditions of mourning and with the period of darkness before the new moon.

The announcement of the *mwami's* burial begins a period of mourning during which fields go unattended, cultivation is ceased, and weeds are allowed to grow. In short, gardens revert to bush: ecological time is reversed. In addition, no one washes or cuts his or her hair, compounds are not cleaned, and sexual activity is ceased. The homestead, in short, becomes like the bush. These conditions remain in force for a period of two months, after which cleansing rituals are held and life returns to normal. Symbolically, however, the period of mourning may last beyond this two-month period, being ultimately terminated with the head-shaving ceremony that occurs when the new *mwami* leaves the *vighala*. This event may not occur for several years, and while the mourning prohibitions do not continue during this extended period, the interregnum, as a whole, is associated, through the symbolism of the head-shaving ceremony, with the conditions of mourning, i.e., the invasion of the homestead by forces of the bush.

The same message is expressed through the symbolic association of the interregnum with the period of darknesss that occurs between the disappearance of the moon's last quarter and the appearance of the new moon. The Bashu call this phase *kavuloyi,* which literally means the 'time of sorcery.' Sorcerers are viewed as being diametrically opposed to the interests of Bashu society, not simply because they cause harm to others but because their actions are antithetical to the communal values of Bashu society. A sorcerer is someone who hordes his wealth and does not share it with others. He eats alone, operates at night, does not attend communal rituals such as weddings and funerals, and sacrifices alone. He represents the individual opposed to the communal interests of society. He is thus a threat to society, which depends on communal action. The association of sorcery with the interregnum period reflects the dangers inherent in the decentralized use of ritual power as well as the chaotic conditions that are thought to accompany interregna.

The dominant figure during the interregnum period is the *mombo,* who, as a young girl, not yet pubescent, is given to the *mwami* at his investiture. She is subsequently viewed by society as the 'queen mother,' who must produce the *mwami's* heir. She is above all, however, the *mwami's* female counterpart, the female side of *bwami,* and represents *bwami* during the interregnum. At the *mwami's* death the *mombo* receives the *mbita* and the *eyisagho y'ovuysano,* containing the *mikene* and pieces of the jawbones of pervious *bami.* She thus takes possession of *bwami,* which she keeps until the new *mwami* is invested.

To understand the significance of the *mombo*'s role, we must turn to the problem of the cosmological position of women in Bashu society.[24] Women are, in a sense, liminal figures in Bashu cosmology. While they are members of the homestead and, in fact, essential to its continued existence, they have strong associations with the world of the bush. These associations give women a certain degree of influence over forces of the bush, allowing them to control wild animals and deflect violent storms.[25] Yet, if women have influence over the forces of the bush, they are also susceptible to possession by these forces and thus a potential threat to the well-being of the homestead. Women are frequently possessed by spirits of the bush that cause sickness and death in the homestead. Their association with the bush is also said to be reflected in their emotional and erratic behavior and their general moral weakness. While restrictions and prohibitions are placed on their behavior—to prevent these associations from outweighing their responsibilities to the homestead—their association with the bush cannot be totally eliminated. Thus women remain liminal figures in Bashu society.

The ambivalent position of women helps explain the position of the *mombo* in the life cycle of *bwami*. Just as women in general ensure the continuity of the homestead through their reproductive roles and their influence over forces of the bush, the *mombo* ensures the continuity of *bwami* by keeping it and guarding it during the period of chaos brought on by the interregnum, and by symbolically giving birth to the new *mwami*. I say symbolically, for the *mombo* was never, as far as can be determined from specific case histories, the biological mother of the new *mwami*, a role that was precluded by her not having reached purberty at the time of the *mwami*'s investiture, and by the prohibition against her subsequently seeing or having sexual relations with the *mwami*. She is, however, regarded socially as the 'queen mother,' and at the investiture of the new *mwami*, she receives him on her lap, designating him as the proper successor and herself as his "mother." In other words, while the *mombo* cannot give birth to the *mwami*, she is responsible for the rebirth of *bwami*.[26]

While the *mombo* contributes to the continuity and renewal of chiefship, her susceptibility to the forces of the bush is said to endanger and weaken *bwami*. The dominance of the *mombo* during the interregnum, therefore, emphasizes the declining condition of *bwami* during this period.

With the investiture of the new *mwami*, *bwami* reenters the world of men and is purified and reborn. So too the land is cleansed, rituals are centralized, the flow of ecological time is renewed, and there is a return of plenty. The cycle of *bwami* is completed and begins anew.

Bashu royal rituals of accession and death provide a window into Bashu thinking about chiefship and its relationship to society. Through them we have seen how the *mwami* is viewed as the primary mediator between the world of the homestead and the forces of the bush. Through his presence and periodic ritual activities, the *mwami* consolidates ritual control over the land, which is necessary for the domestication of natural forces. He also separates untamed forces of the bush from the homestead by periodically taking on the "filth of the nation." In this way, his presence insures the continued productivity and well-being of society, while his death brings on disorder and famine.

THE HISTORICAL CONTINUITY
OF BASHU CHIEFSHIP

Having examined Bashu ideas concerning the nature and function of chiefship as presented by present-day informants, there remains the problem of determining the degree to which these ideas were current during the nineteenth and early twentieth centuries. While the absence of contemporary written records makes this task difficult, there is indirect evidence that attests to the historicity of these ideas.

The clearest indication that the concepts described in this chapter existed during at least part of the precolonial era is the close association between ritual and political leadership exhibited in these concepts. This association is no longer reflected in the actual institution of chiefship, for there has been a gradual separation of political and ritual authority among the Bashu for most of the colonial period. Kitawiti, the first chief to be recognized as *grand chef* by the Belgians in 1923, was viewed by the Bashu as possessing both ritual and political authority. He died, however, after only six months in office, and none of his successors, with the possible exception of his brother Muhashu, who ruled from 1935 to 1945, was viewed as possessing any ritual authority over the land. Since 1945 the two forms of authority have followed separate lines of succession. This de facto separation of ritual and political authority suggests that their merger in Bashu political thought must have developed at an earlier time, prior to the colonial era.[27]

This conclusion is supported by ethnographic evidence on the ritual and political practices of former chiefly families who preceded the present chiefs into the Bashu region. This evidence indicates that many of the elements of chiefship discussed in this chapter existed among the Bashu prior to the arrival of the present chief's ancestors at the beginning of the nineteenth century. To begin with, data on alliances between these former chiefs and local ritual leaders at the time of the present chiefs' arrival suggest that the coordination of local ritual authority over the land had already become politically important among the Bashu. These alliances will be discussed in chapter four. Second, similarities between the ritual practices of present chiefs and those of earlier chiefly families —the later practices preserved as a result of the continued ritual importance of these families in the performance of land rituals and by their incorporation into the ritual structures of chiefship—indicate that the practices described in this chapter have existed among the Bashu for a considerable time. These similarities as well as significant differences are listed in Table 2.

It is, of course, possible that the practices of former chiefly families have been contaminated through contact with those of the present chiefs, and may not, therefore, represent independent sources of evidence. The distribution of many of these practices in the wider Lakes Plateau Region, however, supports the antiquity of these practices and suggests that if contamination has occurred, it has been in the direction of the present chiefs' practices rather than toward the practices of earlier chiefs.

In Table 2 I have noted the distribution of political practices and terminology among the Bashu and in other Western Lakes Plateau societies. In order to illustrate the geographical distribution of Bashu practices, I have arranged these societies roughly according to their geographical relationship to the Bashu. Where a specific practice has a limited distribution but seems to be related structurally and functionally to practices found in other societies, these additional locations are marked with a slash. Where the ethnographic record is unclear as to whether a practice exists in a particular society, I have indicated this with a question mark.[28]

Table 2, while hampered by gaps in the available ethnographic record, indicates that in terms of their political culture the Bashu are more closely associated with the states of the lake region of Zaire—southern Nande, Hunde, Nyanga, Havu and Shi—than with the Babito states of western Uganda, from which their present chiefs have come. This in turn suggests that the bulk of Bashu political culture either existed in Isale before the present chiefs arrived or was borrowed from the south after their arrival.

Bashu historical traditions suggest that both processes occurred. When the present chiefs settled in Isale, they were initially invested by a family of preexisting rainchiefs, who provided them with some of their ritual regalia and instructed them in the performance of royal rituals related to chiefship. Subsequently, the senior branch of this ruling family found it necessary to recruit other ritual specialists to direct their investiture

TABLE 2
THE DISTRIBUTION OF ROYAL RITUAL PRACTICES IN THE
WESTERN LAKES PLATEAU REGION

	Shi	Rwanda	Havu	Nyanga	Hunde	S. Nande	Bashu-Babito	Bashu	Konjo	Nyoro	Nkore	Ganda
Ritual burial of *mwami*				X		X	X					
Ritual banging of hammers by *mwamihesi* during investiture			X	X	?	X	X					
Mwami smeared with red powder				X	X	X	X	X				
Mwami receives emblems of authority on *mukako*				?	X	X	X	X				
Mombo as female counterpart of *mwami* and 'mother' of his successor				X	X	X	X	X				
Mwami protected by *muhangami**				X	X	X	X	X				
Royal diadem made of wild plants and teeth of savage animals	/	/				X	X	X				
Role of original forest dwellers as keepers of royal fire	X	X	?	X	X	X	X	X				
Mwami's lower jawbone removed at his funeral				/	/	/	/	X	X	X	X	/
Burial of *mwami* in wooden log associated with a canoe			X		X		/	X	X	X		
Prohibition against use of iron during *mwami*'s funeral			X	X	X	X	X	X	X	X	X	
Prohibition against sexual relations during investiture			X	X	X	X	X	X	X		X	
Royal compound called *kikali*							X		X	X		
New moon ceremonies to renew chiefship		?					X		X	X	X	

*In Kihunde and Kinyanga the term is *muhakabi.*

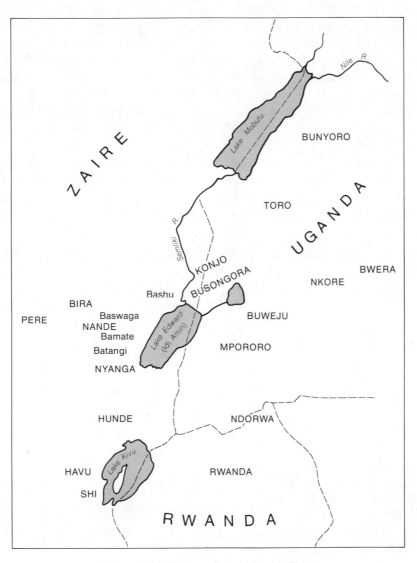

MAP 1. The Western Lakes Plateau Region

ceremonies. These new ritual leaders came from among the Batangi of Musindi in the southern Nande region. The recruitment of these new leaders apparently led to the introduction of other ritual elements into Bashu political culture and to the cultural differentiation of this senior branch of Bashu chiefs from collateral branches and from earlier rain-chiefs (see chapter 4 for details of these events). A comparison of the distribution of southern Nande practices among the Bashu and their Nande and Konjo neighbors to the north and east of Isale supports these traditions and suggests which elements of Bashu political culture are recent innovations and which are of greater antiquity.

Table 3 shows that the ritual burial of the *mwami* in the *vighala* and the banging of hammers by the *mwamihesi* are found among the southern Nande and the senior line of Bashu chiefs, but not among other ruling lines or among neighboring groups to the north and east. The use of red

TABLE 3

THE DISTRIBUTION OF SOUTHERN NANDE ROYAL RITUAL PRACTICES
AMONG THE NORTHERN NANDE

	Senior line of Bashu chiefs	Other Bashu ruling lines	Konjo of western Ruwenzoris	Banisanza of Lisasa
Ritual burial of *mwami*	X			
Banging of hammers at investiture	X			
Mombo	X	X		X
Prohibition against burying *mwami* in earth	X	X	?	X
Prohibition against use of iron and sexual relations during funeral and accession of *mwami*	X	X	X	X
Mwami smeared with red powder at investiture	X	X		
Role of *muhangami*	X	X	X	
Mwami invested on *mukako*	X	X	X	X
Mbita, royal diadem, made from wild plants and animals	X	X		X

powder occurs among all Bashu chiefly lines but apparently not among the Banisanza or Konjo. The other elements occur in each of the areas examined. This distribution suggests that the banging of the hammers and the *vighala* may be more recent innovations, perhaps introduced following the recruitment of royal investors from among the southern Nande. The fact that these two practices are related functionally in the accession ceremony, with the banging of hammers occurring as the *mwami* leaves the *vighala,* suggests that they were borrowed together. The other elements listed in the table, which have a wider distribution, are probably of greater antiquity and may have existed among the Bashu before the present chiefs arrived. The distribution of ritual practices within the Western Lakes Plateau region, therefore, supports internal ethnographic evidence from among the Bashu that indicates that Bashu political culture predates the establishment of the present line of Bashu chiefs.

Finally, it must be noted that there is a fundamental consistency between the Bashu's political ideas and their general view of the world in which they live. Chiefship is, in fact, a product of this wider world view, and specifically of ideas about the relationship between the homestead and the bush, the sources of misfortune, and the need for corporate as opposed to individual action. In this sense, chiefship is simply a variant of Bashu ideas about healers, sorcerers, and women. Similarly, the symbols and ritual actions associated with chiefship are by no means unique to chiefship but are variants of symbols and rituals found throughout Bashu ritual life, and particularly in conjunction with the major life-crisis rituals of birth, circumcision, marriage, and death.[29] This internal homogeneity is consistent with evidence supplied by ethnographic and distributional data that suggest that Bashu chiefship is of considerable antiquity. At the very least this consistency makes it unlikely that chiefship is a recent innovation grafted onto Bashu culture from the outside.

There is reason to believe, therefore, that the political ideas contained in the testimonies of present-day informants existed in more or less the same form during earlier periods of Bashu history going back to perhaps the beginning of the nineteenth century or even earlier. This is not to suggest that Bashu political ideas were static, for, as will be seen in the following chapters, changes did occur in the way in which chiefship was perceived during the period in question. Yet these changes can best be seen as elaborations rather than radical transformations in the constellation of ideas surrounding chiefship. The central ideas concerning the relationship between the world of the homestead and the forces of the bush, the need for corporate ritual action, and the central position of the *mwami* in coordinating ritual and in maintaining the well-being of society have undoubtedly exhibited a remarkable continuity over long periods of time, a conclusion that is supported by their persistence in the face of rapid and deep-seated social and economic change during the colonial and postcolonial eras.

While it is possible to show that Bashu ideas about the nature and function of chiefship are of considerable antiquity, it is more difficult to prove that these ideas actually affected the ways in which people acted in particular historical situations. Some may even suggest that it is impossible to do so. Thus Vansina (1978a: 210), discussing the Kuba of Zaire, has concluded that, "We can only guess at the kind of motives that drove Kuba leaders and followers to action, and we must realize that we are groping for shadowy patterns in the absence of evidence about personalities." Nonetheless, there is a remarkable correlation between the ideas described in this chapter and patterns of political competition and development during the nineteenth and early twentieth centuries. In Part Three of this study, I will argue that this correlation is not coincidental but reflects the impact of Bashu cosmology on political perceptions and actions. The intervening chapters will set the historical and ecological stage for this exploration of the politics of ritual chiefship.

PART TWO

THE FOUNDATIONS
OF BASHU CHIEFSHIP

CHAPTER TWO

MIGRATION, SETTLEMENT, AND THE ROOTS OF BASHU CHIEFSHIP

ECOLOGY AND SETTLEMENT IN THE UPPER SEMLIKI VALLEY REGION

From their homes in the Mitumba Mountains the Bashu look out daily over the rolling plains of the Semliki Valley stretched out a thousand meters below, and beyond the valley to the western slopes of the Ruwenzori Mountains. On a clear day, such as occurs after the rains in March and April, it is possible to see the snow-capped peaks of the Ruwenzoris and, beyond the southern reaches of these mountains, the broad grasslands of western Uganda.

To understand the history of the Bashu chiefdoms it is necessary to place it within the historical and ecological context of this wider geographical region. For the history of the Bashu region is one of the ascent of successive groups from the grasslands of the Rift Valley up onto the slopes of the mountains, and of the interaction of these groups.

The Mitumba Mountains rise sharply from the floor of the Semliki Valley to a mountain plateau, the average elevation of which is 2000 meters. This escarpment is transected at several points by river valleys, which provide channels of communication between the plains and the highlands. From the end of the seventeenth century, or perhaps earlier, Nande agriculturalists expanded their seed cultivation and settlements along these valleys and came into contact with forest dwellers who had preceded them into the mountains. Later, as the Nande cleared away the forest cover, these same valleys provided the paths by which pastoral groups expanded their political influence into the mountains and interacted with the agriculturalists. Toward the end of the nineteenth century, the river valleys of the Mitumbas provided avenues by which Manyema ivory traders, the armies of Bunyoro, Rwandan irregulars under the leadership of Karakwenzi, and African agents of Belgian colonial rule penetrated the mountain highlands. These groups raided extensively throughout the Bashu region and, like previous intruders, affected the political development of the Bashu chiefdoms. Finally, it was along these same valleys that Belgian administrators initially expanded their grid of political control beyond the Semliki Valley during the first decades of the twentieth century.

The most important break in the Rift wall occurs in Isale, where a number of river valleys converge to form a major gap in the mountains. The heart of this network is the valley of the Talia River, which runs northward from its source in the southern Mitumbas, winding its way through the higher mountains before turning eastward as it passes through Isale and descends to the Rift floor. Just before the Talia enters the plains it is joined by several tributaries descending from the west, south, and northwest. The convergence of these valleys, dividing the Rift wall into a number of intersecting mountain spurs, made Isale a major focus for migration, settlement, and political development within the mountains. A second major center of migration and settlement lay to the south of Isale along the valley of the Tumbwe River, which descends from the mountains into the plains near the northwest corner of Lake Edward.

Until the end of the seventeenth century the northern Mitumba Mountains were covered with a thick mountain forest, similar to that which can still be seen on the upper reaches of Mounts Muleke, Kasongwere, and Kyavirimu and can be found in abundance south of Lubero.[1] The Bashu claim that the forest was occupied by the Basumba, whom they equate with the Bambuti Pygmies of the Ituri Forest. The term Basumba, however, does not refer exclusively to Pygmies, but means 'those who live in the forest,' as opposed to people like the Bashu, who clear the forest in order to plant their crops. It thus refers to forest Bantu groups such as the Bapere, Bapakombe, Babira and Baamba, as well as to the Bambuti.[2] The traditions of the few groups of Basumba who remain among the Bashu, as well as those of the Bapere and Bapakombe, who claim to have formerly inhabited the lower regions of the Mitumbas,[3] suggest that the Basumba referred to in these traditions were forest agriculturalists rather than Pygmies.

The western slopes of the Ruwenzori Mountains may have been occupied by a similar population of forest dwellers, for the equatorial forest stretched across the Semliki Valley, as it still does north of Beni, providing a corridor by which forest peoples could have expanded their settlements onto the Ruwenzoris. Thus, the Baamba, who presently occupy the forested areas around the north end of the Ruwenzoris, are said to be related to the Babira of the Ituri Forest.[4] The existence of forest agriculturalists on both sides of the Semliki Valley is supported by a tradition

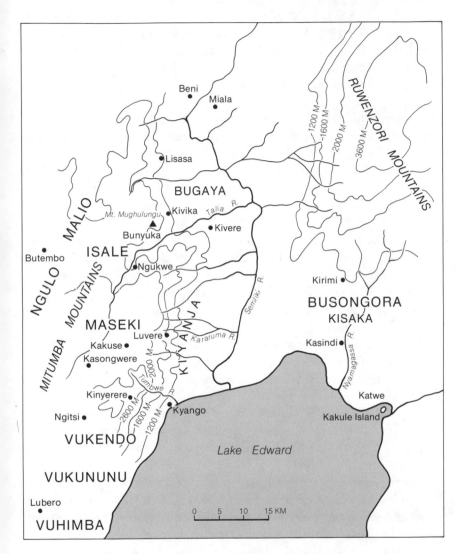

MAP 2. The Upper Semliki Valley Region

collected among the Bambuba peoples who presently live in the forested region of the Semliki Valley north of Beni. This tradition describes how the Bambuba travelled south from the region around Irumu and Epulu, settling on the slopes of the Ruwenzoris and later in Isale, only to be subsequently driven off by the arrival of the Bashu and Banisanza (a Nande group that presently occupies the western slopes of the Ruwenzoris).[5]

It is possible that pygmanoid groups related to the present Bambuti preceded, or even coexisted, with these forest cultivators in certain regions of the Mitumba Mountains, as well as on the slopes of the Ruwenzoris. Father Leiven Bergmans (1970: 11), who has studied Nande culture and history for a number of years, states that the Bambuti almost certainly lived as far east as the Nile-Congo watershed, which runs along the crest of the Mitumbas, but probably not to the east of this line. He notes as evidence references to Pygmies in the stories and proverbs of the Nande living to the west of this crest, but the absence of such references in the stories and proverbs of those living to the east of it. On the other hand, evidence from Heinzelin's excavations at Ishango at the mouth of the Semliki River suggests that a hunting and fishing population with late mesolithic industries preceded Bantu settlement in the area and had contact with the Bantu occupants (1957:77). He further notes that local traditions refer to those former occupants as being of short stature. While similar traditions are found throughout much of East Africa, and may be culturally determined, they may also indicate that a Bambuti-related group did exist in the Semliki Valley when the forest cover was more pervasive. There are, of course, pygmanoid peoples living around the northern slopes of the Ruwenzoris today, though these people may have migrated from the north.

The Nande/Konjo agriculturalists who presently occupy the Mitumbas and Ruwenzoris claim to have moved into the mountains between nine and eleven generations ago, and thus from the end of the seventeenth to the beginning of the eighteenth century, though this movement may have begun earlier. Swaga traditions collected by Bergmans (1970: 26–31) indicate that the movement began between thirteen and fifteen generations ago, and thus around the end of the sixteenth century (see Appendix I). Among these early settlers were members of the Bito (not to be confused with the Babito, who arrived later), the Bakira, Bahombo, Bahera, Baswaga, Batangi, and Bashu clans.

The original homelands of these groups are difficult to determine, though there is reason to suspect that some of the earliest settlers were

Sudanic speakers who migrated from the northwest, and that these Sudanic speakers were later joined by Bantu groups moving in from the
west and then later from the east.[6] The traditions of both groups are
unanimous in claiming that their ancestors occupied the grasslands of the
Western Rift Valley for several generations before moving into the
mountains.

While the Semliki Valley is presently a national park and was dominated by pastoralists when the first Europeans arrived in the 1880s and
1890s, traditions collected at the fisheries located at Kyavanyonge on the
northern shore of Lake Edward indicate that the valley was once a center
of rich cultivation.[7] The presence of grindstones for millet and other
grains in abandoned settlements and in the upper levels of the archeological excavation at Ishango appears to support these traditions. The inhabitants of these settlements, located along the lake shore and on the banks
of the Semliki and its tributaries, evidently combined seed cultivation
with fishing and the herding of goats, sheep, and perhaps some short-
horned cattle.[8]

It is difficult, if not impossible, to determine just how long these groups
occupied the grasslands before expanding their settlements into the
mountains. Roland Oliver (1954: 31–33), following Johnston (1904) and
Stanley (1890), suggests that the present Nande/Konjo population represents the remnants of the original agricultural population of the lakes
plateau, and that they have occupied the grasslands of the Western Rift
Valley from very early times. Heinzelin (1957: 76) suggests that the
agricultural settlements at Ishango date from before A.D. 1000, while
data from western Uganda place the arrival of Sudanic-speaking groups
during the eighth century and the establishment of the earliest Bantu
settlements between the ninth and tenth centuries A.D. (Buchanan,
1974: 69).

A more critical question is why these plains dwellers chose to abandon
their settlements in the Semliki Valley and move into the cooler and
wetter region of the mountains. There is, after all, a natural tendency for
people whose economy is based on the land, and on their knowledge of
it, to resist moving into a new physical environment. Stanley (1890: 283)
hypothesized that the Konjo/Nande people retreated into the mountains in response to the invasion of their homelands by culturally and
militarily superior Hamitic peoples, who were the ancestors of the Hima
herdsmen who began settling in the Semliki Valley perhaps as early as the
thirteenth century (Buchanan, 1974) and dominated the region when
Stanley arrived. This variant of the Hamitic myth was no doubt encour-

aged by the fact that both the Konjo and Nande were at the time being subjected to constant raiding by the armies of Bunyoro. A group of Bakonjo, in fact, sought Stanley's assistance in ridding their country of these invaders (Stanley, 1890: 284). There may, nonetheless, be some truth to this hypothesis, although it needs to be examined in light of the changing physical environment of the upper Semliki Valley region between the sixteenth and the eighteenth centuries to be understood fully.

Recent work on the climatological history of the lakes region of East Africa by Cohen (1974), Riehl and Meitin (1979), and Herring (1979) suggests that the migration periods cited in Nande tradition—the middle of the sixteenth to the middle of the seventeenth and the end of the seventeenth to the middle of the eighteenth century—were marked by periods of extended dry weather, which led to drought conditions and famine for the agricultural peoples of the region. These conditions evidently caused people throughout the region to look for wetter lands near lakes, rivers, forests, and mountains and may well have contributed to the movement of Nande cultivators out of the Semliki Valley. It should be noted, however, that the approximate nature of Bashu chronology makes it difficult to make a precise correlation between these periods of drought and Nande migrations. On the other hand, evidence from more recent periods suggests that drought conditions could have been an important stimulus to migration.

The impact of drought conditions on the environment of the upper Semliki Valley was described by Stanley, who visited the region during a period of relatively low rainfall at the end of the nineteenth century:

> We were in a much drier climate and the superficial aspect of the country was much as might be expected from a comparatively rainless district—it was of a worn out and scorched country. The grass was void of succulency and nutriment. The slopes of the rounded hills presented grooves of a brick dust colour; here and there grew a stunted tree with wrinkled and distorted branches and ugly olive green leaves, too surely denoting that the best of the soil had been scoured away or consumed by annual conflagrations, that vegetable life was derived under precarious circumstances despite copious showers of the rainy season. (Stanley, 1890: 223–24.)

Such conditions would certainly have made cultivation difficult. Lugard (1893: 175–78), who travelled more extensively in the region, presented a similar picture, describing most of the area as a dry grass plain stunted with euphorbia and acacia. He noted, however, that extensive cultivation occurred along the mountain edge, in and around the river beds where the Talia and other tributaries joined the Semliki, along the lake shore, and further north where Nande settlements extended several miles into the forest. Lugard's observations were confirmed by Stulhman (1894: 279–84), who crossed the valley with Emin Pasha in 1891. This picture conforms to the general pattern of population movements during periods of drought suggested by Herring (1979) and Webster (1979) and supports the suggestion that drought conditions in the valley could have caused Nande cultivators to move out of the area.

If drought and famine did play a role in stimulating Nande migrations, they may not have worked alone. As E. Steinhart (1979) has shown in his study of the pastoral kingdoms of Uganda, drought conditions favored herdsmen over agriculturalists and could lead to the political domination of the cultivators by the herdsmen. This was because the herdsmen's cattle could survive the drought conditions that destroyed the cultivators' crops. Under such conditions, farmers ". . . reduced to desperate straits by famine might well accept service among pastoral neighbors with a stock of valuable and mobile resources such as cattle" (p. 203).

There is evidence that this process occurred in the upper Semliki Valley in response to the drought conditions that hit the region at the beginning of the eighteenth century (Herring, 1979). For it was at this time that Bahima-Babito chiefs, who were subjects of the Babito king of Bunyoro and who had coexisted with Nande cultivators for several generations, consolidated their political authority over Busongora to the east of the Semliki River (Rukidi III, n.d., p. 7).[9] In the context of increased competition between herdsmen and cultivators over deteriorating water and land resources, cultivators who preferred to resist domination by the herdsmen and to maintain their agricultural existence may have chosen to move away from the centers of Bahima control. In this way, conflicts between herdsmen and cultivators may have contributed, as Stanley suggested, to the movement of the Nande and Konjo groups out of the Semliki Valley.

Disease may also have combined with drought conditions to cause Nande and Konjo groups to settle in the mountains. There are a few scattered references in Nande traditions to disease having caused people to move out of the plains. Bergmans presents two such traditions, one from the Bashu, the other from the Baswaga. The Bashu tradition refers to the arrival of the ancestors of the present chiefs around the beginning of the nineteenth century. It states that many people were dying in the plains and that this caused the survivors to move into the mountains. The disease is not named. However, Bergmans (1974: 8) concludes, without explaining why, that it may have been malaria. The Baswaga tradition (Bergmans, 1970: 18) refers to an earlier migration that is said to have occurred between thirteen and fifteen generations ago, and thus roughly from the end of the fifteenth to the middle of the sixteenth century. This tradition refers explicitly to sleeping sickness. To the south of the Bashu region, Batangi traditions speak of sleeping sickness having driven people out of the plains to the south of Lake Edward around this same time. My own research among the Bashu revealed only two references to disease as a stimulus for migration, both from the descendants of early Nande settlers in the Bashu region. Only one of these specifies the disease, sleeping sickness.[10]

It is difficult to draw any definite conclusions about the role of disease as a stimulus for migration from these few scattered traditions. The references to sleeping sickness north of the lake may represent the incorporation of recent experiences with the devastating sleeping sickness epidemic that broke out in the valley at the beginning of the twentieth century and with the subsequent evacuation of the valley, into traditions that refer to earlier events. On the other hand, Webster (1979) indicates that sleeping sickness spread into Busongora from Bunyoro during the major drought that occurred at the end of the sixteenth and beginning of the seventeenth century. John Ford (1971), moreover, has suggested that sleeping sickness may have been endemic in the Semliki Valley before the outbreak of the 1905 epidemic. He supports this suggestion with the following tradition, told to Dr. van Hoof by the Belgian Territorial Administrator M. Hackaers in 1926. The tradition indicates that a succession of people settled in the Semliki Valley only to be driven out by outbreaks of sleeping sickness.

> The Mabudu, for example, who live in the Nepoka, were settled over 200 years ago in the Semliki, to the north of Ruwenzori; they abandoned that part of the country, so they say, to escape the ravages of sleeping sickness.
> Similarly the Bakumu or Babira, whose vanguard, coming from the east, passed around the south of Ruwenzori and then spread northwards and westwards, to the district of Uvumu and Mombasa, only remained for a short time in the Semliki Valley where the dreadful disease killed more of their people than the wars they had had to wage in order to force their way across the high plateau dominating the fatal valley (Ford, 1971: 177).

The epidemiology of sleeping sickness corresponds to the changing environmental conditions of the upper Semliki Valley described above. Ford (1971: 182) states that sleeping sickness may change from an endemic to an epidemic disease when the ". . . physiologically and ecologically balanced system, comprising human infective trypanosomes, riverine tsetse, and man, is broken down and involved with the spread of other tsetses dependent primarily upon wild animals. . . ." In other words, sleeping sickness may exist as a mild endemic disease where concentrated settlements and well-maintained fields separate man from the reservoir of trypanosomes and tsetse dependent primarily on wild animals. When settlements begin to break up, allowing the bush to encroach upon the remaining homesteads, these wild animal tsetse become vectors for the transfer of trypanosomes between the wild animal reservoir and man. The result is an increase in the incidence of sleeping sickness to epidemic proportions. Such a break-up of settlements may have occurred among the riverain population of the Semliki Valley as a result of drought conditions and declining agricultural productivity during the seventeenth and eighteenth centuries. As sections of these settlements began to move off toward the mountains in search of better lands, the areas they had abandoned were reclaimed by the bush; settlements that remained became more scattered; and the ecological balance that had kept the incidence of sleeping sickness at endemic levels broke down. Following Ford's hypothesis, this could have resulted in an outbreak of sleeping sickness among the agriculturalists who remained in the riverain settlements, forcing them to follow their kinsmen towards the mountains. Thus sleeping sickness may have been a secondary stimulus for migration.[11] Yet, if this is so, how were the pastoral groups that stayed in the valley at this time able to avoid contracting the disease? The answer may be that the pastoralists simply avoided the infected riverain areas. In fact, pastoral residents of the Semliki Valley told Belgian officials charged with evacuating the valley in 1932 that they had no fear of the disease because they avoided areas in which the disease was prevalent.[12]

The movement of cultivators into the mountains need not have been permanent, and it is possible that with the return of more favorable climatic conditions some farmers may have returned to the valley. This pattern of temporary dislocations has, in fact, marked more recent periods of famine among the Bashu. On the other hand, repeated periods of dry weather, such as evidently occurred during most of the eighteenth century (Herring, 1979: 58), and perhaps disease, combined with gradual acclimation to the more productive mountain environment, may have encouraged a growing number of agricultural settlers to establish permanent residences in the Mitumbas and Ruwenzoris. This decision was no doubt reinforced during the eighteenth century by the arrival of many new herdsmen from western Uganda. As will be seen in chapter three, the arrival of these pastoral immigrants shifted, once and for all, the political balance between herdsmen and cultivators in the valley in favor of the herdsmen. While some cultivators did settle in the valley on into the nineteenth century, in most cases they did so specifically to become clients of the herdsmen, in order to acquire goats for bridewealth payments. By the nineteenth century, cultivation within the valley, even during periods of relatively high rainfall, was limited to a few settlements along the rivers and lake shore, which were under Hima political domination.

SETTLEMENT AND CULTURAL INTERACTION IN THE MITUMBA MOUNTAINS

The Nande cultivators who moved into the Mitumbas evidently settled first in the river valleys that descend from the mountains, clearing away the forest cover and adjusting to the new environment. Later they expanded their settlements into the upper regions of the mountains. The Isale region, because it offered easy access to the mountain plateau, attracted a large number of immigrants and became a major settlement area from which groups expanded into neighboring regions. This pattern of settlement is indicated by the location of senior clan settlements within the mountains. Thus, members of the Bahera clan living to the south of Isale in Maseki, Kakuse, and Kasongweri, and to the north of Isale in Mulali and Vutongo, trace their origins to the Bahera settlement in Ngukwe, a hill that rises from the Talia river in Isale, and before that to Vurarama in the valley of the Talia. Similarly, the Bakira claim to have expanded their settlements from two initial locations in Isale, Vusereghenya, located just north of the Talia, and Kamuteve, south of the Talia.

Other Nande groups moved up into the mountains via the valley of the Tumbwe River to the south of Isale. These migrants spread westward into Buswaga and north into the southern Bashu region. Still others moved south along the lake shore and up other river valleys descending from the mountains.

The movement of Nande agriculturalists into the Mitumba Mountains brought them into contact with forest cultivators who had preceded them into the region. The Bashu claim to have chased away these forest peoples, though evidently not without some resistance, since there are traditions that refer to "Basumba" raids on Bashu settlements. In fact, as late as the beginning of the present century, the Mwami Lukanda of Maseki is reported to have complained of Basumba raids to a Belgian reconnaissance party.[13] However, more recent Nande expansion into the forested regions of the Babira, Bapere, and Bapakombe, along with evidence of extensive cultural interaction between the Nande and these forest peoples, suggests that these traditions oversimplify the process of Nande expansion, and that it involved more than armed confrontation.

Schebesta (1936: 156), for example, described Nande expansion north of Beni in the 1930s as a peaceful process:

> Bungulu lies on the southeast fringe of the forest, and is exclusively inhabited by Bananda [Nande] who are steadily encroaching more and more on those primeval woodlands. The Bananda are folk of the steppes and loathe the forest, which they are constantly clearing to make new settlements. The Babira give way unresistingly before the advance of the enterprising Bananda. Only in one village, the very last one which the Bananda occupied, did I notice that its chief was a Babira man. From this circumstance one infers that the Bananda are following a process of peaceful penetration of the forest zone. Indeed I have noticed that a sort of mixed culture has evolved from the commingling of the inhabitants of the forest and the steppes. This was apparent in the various types of huts which I saw in this region.
>
> The Bananda huts were circular and thatched with straw while the Babira have gabled-roof huts thatched with leaves. It was not until I reached the river Biena that I saw the first genuine Babira village. . . .

Schebesta's description, while perhaps overestimating the dominance of the Nande, indicates some of the main features of Nande expansion. The first and perhaps most important of these was the Nande practice of clearing the forest. By doing so the Nande changed the physical environment of the region into which they expanded, making it less suitable for the type of forest agriculture that the Basumba practiced and more suited to the cultivation of their own seed crops. Clearing the forest also drove away the game and therefore harmed the economy of the forest people. Some of the original inhabitants were thus driven away from the seed-agriculture fringes and retreated further into the forest in order to grow their crops and catch game, while others, perhaps a significant number, were evidently assimilated into Nande society. Schebesta's description also indicates that the Nande did not advance along a single line, mowing down the forest as they went. Instead they infiltrated the forest and settled among the Basumba in small groups, recognizing the political authority of the Basumba in whose villages they settled. Thus Schebesta notes the presence of Nande-styled huts deep in the forest and the existence of a Mubira chief in the most recent area of Nande settlement.

Evidence from recent Nande expansion among the Bapere suggests that the Nande initially settled among the forest dwellers as traders, exchanging their more abundant agricultural produce, and perhaps salt from Katwe and pottery, for meat, resins, red powder, raffia, and honey. Having been accepted within a village as a trader, the newcomer would be joined by his relatives and others, who took up farming and began clearing the land. This resulted in some of the forest dwellers moving deeper into the forest, while at the same time, the number of Nande in the village grew. Eventually, the Nande represented a majority of the population in what had formerly been a forest village.[14]

The Nande system of land tenure may have facilitated their expansion. The Bapakombe and Bapere claim that prior to Nande expansion land was a free commodity, to be used by anyone who chose to cultivate it, without obligation to the person who had cleared it. When the Nande arrived they introduced a system of land tenure in which a person who settled on a piece of land that had previously been cultivated was ritually, and thus to a certain extent politically, dependent on the clearer of the land. This system led to the development of patron-client relationships between first occupants (bakonde) and later settlers. The emergence of these relationships may have given the Nande settlers an organizational advantage over their forest hosts by providing a means for mobilizing support on a territorial basis.

The slow infiltration of seed-agriculturalists among the forest Bantu clearly led to considerable cultural borrowing between these groups. The most important element borrowed by the Nande was the men's clubhouse (kyaghanda). Among the forest Bantu this hut was located in the center of the village, which itself consisted of a street or double row of houses (Wachsmann and Trowell, 1953: 9). Within the clubhouse the men of the village met to discuss important events and to socialize. The Nande borrowed the clubhouse idea and adapted it to their own pattern of dispersed settlement. Each localized clan section, nda, possessed a kya-ghanda where its members met along with nonagnatic kin and others who may have settled among them and received land. The *kyaghanda* was evidently a center for cultural exchanges between the Nande and the Basumba, for the Bapere claim to have borrowed the beer-drinking organization used in the clubhouse from the Nande. They also borrowed a number of political practices associated with chiefship and land tenure.[15] Other items that the Nande appear to have borrowed from the people of the forest include: bow and arrow styles, the heavy iron neck rings worn by Nande women, raffia arm rings *(vutegha)*, and certain knife styles. It is also important to note that the Nande recognized the ritual authority that the Basumba had over the forest and therefore incorporated the Basumba into their ritual organization, even though they claim to have chased them all away in their traditions of settlement. It is the descendants of Basumba who are responsible for killing the ritually dangerous flying squirrel in order to obtain its skin for making the *mbita*. A Musumba must also relight the royal fires following the investiture of a new *mwami*.

THE ORGANIZATION OF MOUNTAIN SETTLEMENTS AND THE ORIGINS OF CHIEFSHIP

Early Nande settlements were organized along lines of agnatic descent, with the members of a single minimal patrilineage *(nda)* clearing and occupying all, or part, of a mountain ridge, their homesteads separated from one another by banana plantations and other gardens. Beyond these localized lineage settlements, however, descent appears to have played a minimal role in regulating social interaction. Today, Nande clans, though named groups, have very little corporate identity. They are not geographically cohesive, nor are they totemic or exogamous units. Any two sections of the clan remain exogamous only as long as they remember their common ancestors; after that, marriage within the clan is acceptable.

In the political-jural domain the members of one localized lineage can call on members of other lineage sections of their clan for assistance or to help settle disputes among lineage members, but they might just as frequently call upon maternal or affinal kin. Similarly, in religious matters, the senior member of each localized lineage, the *mukulu*, is responsible for the performance of ritual offerings to ancestors and other spirits. The members of several lineage sections occasionally come together to participate in a sacrifice to a common ancestor or to attend a marriage or funeral. However, on such occasions the members of each lineage section eat separately, indicating their semiautonomous status.

It is possible that clans played a more important role during the early years of mountain settlement. This is suggested by the presence of joking relationships *(kyavisa)* between Nande clans. While no great importance is presently attached to these relationships, and many younger Nande are unaware of who their joking partners are, older informants claim that formerly, clans that joked together also preferred to intermarry and generally protected one another. These joking partnerships may thus be a vestige of an earlier system of interclan relations and an indication that clan identity was once more important than it is today. The fact that the pairing of clans that joke with one another is the same among the Konjo as it is among the Nande—Baswaga joke with Bito, Bahera with Bakira, Bashu with Bahombo—suggests that these relationships evolved during the period in which the ancestors of both groups lived together in the Western Rift Valley.[16]

If clan membership was formerly more important than it is today, its decline as an organizational principle may in part be a product of the mountain environment into which the Nande moved. The ridges and valleys of the Mitumba Mountains created natural barriers to the maintenance of large cohesive clan settlements and to communication between separate clan sections. In addition, the attraction of the river valleys for early settlers tended to concentrate settlements in limited areas. This in turn encouraged the development of the system, described above, of territorial organization based on land ownership, in which clan sections that first cleared the land attracted a clientele of later arrivals, who were given land in return for recognizing the dominance of the first occupants. The resulting multiclan neighborhoods *(miyi,* sing. *muyi)* as opposed to single clan settlements, thus became the primary unit of social organization.

Descent organization may also have been undermined by the development of early forms of ritual chiefship that cross-cut and subordinated Nande descent groups. As indicated in chapter one, these early forms of leadership were the antecedents of the type of ritual chiefship that emerged during the nineteenth and early twentieth centuries. The most important of these early forms of territorial leadership, given the nature of the mountain environment, was that based on the power of rainmakers.

The authority of Nande rainchiefs was based on their effective use of rainstones: quartz crystals and what appear to have been neolithic digging-stick weights, which their ancestors are said to have found in the ground when they first arrived in the mountains. The quartz crystals were several inches in length and thought to be male, while the digging-stick weights had a hole in the center and were female. Both stones were rubbed with resin oil and placed in the sun so that they became shiny. They were then placed together in a hole that was lined with wild banana leaves and filled with water and a mixture of herbs. The hole was then covered with more wild banana leaves. The combining of the digging-stick weight, associated with the earth, with the quartz crystals, which were said to be solidified rain, was thought to bring rain. Specifically, it attracted rain from the Ruwenzori Mountains, the snow-capped peaks of which were believed to be made of the same substance as the quartz crystals. The local name for the Ruwenzoris, *Runzororo*, is in fact derived from the term *nzororo*, which denotes both the quartz crystals and the snow that caps the Ruwenzoris. It should be noted that from the Mitumbas rain appears to move west from off the Ruwenzoris.[17]

The authority of rainchiefs was thus distinct from that of later Bashu chiefs, in that it was based on the possession of specialized ritual powers rather than on their central position in the coordination of the ritual powers of others. Rainchiefs were ritual specialists who were function-ally and conceptually equivalent to 'healers of the land' and 'clearers of the forest,' and their relationship to these other ritualists may initially have been marked by equality and informal cooperation.

By the beginning of the nineteenth century, however, certain rain-chiefs had achieved a degree of dominance vis-à-vis other ritual leaders, becoming a first among equals and a central focus for the coordination of ritual authority over the land. This transformation was evidently made possible by the central importance that the Bashu attached to controlling rainfall and by the extensive economic resources, in the form of payments for rainmaking, that were consequently available to rainmakers. These resources permitted successful rainchiefs to establish networks of alli-ances with the lineage heads who employed their services and, more importantly, with other ritual leaders. These alliances made rainchiefs a natural focus for coordinating ritual activity. Thus the rainchief Mukirivuli, who controlled the rains in the Mughulungu areas of Isale at the beginning of the nineteenth century, succeeded in establishing alli-ances with several 'clearers of the forest' and 'healers of the land' as well as with other lineage heads in the areas surrounding his home and, accord-ing to the descendants of these allies, directed the performance of land rituals for the area as a whole.[18]

In accordance with their new role as coordinators of ritual authority, rainchiefs evidently began to acquire an ideological association with the general condition of the land. This is reflected here, as it is in the case of later chiefs, by the existence of prohibitions against sex, nursing, and the use of iron hoes, axes, and billhooks during a rainchief's accession and mortuary rites. On the other hand, rainchiefs had apparently not reached the level of identification with the land achieved by later chiefs. Thus, according to the descendants of former rainchiefs who are presently rainmakers, the daily lives of rainchiefs were not as restricted by ritual prohibitions as were those of later chiefs. Those avoidances that did apply were only operative during the period in which the rainchiefs were performing their ritual duties, whereas the lives of later chiefs were continually monitored.

Rainchiefs also differed from later chiefs in regard to the size of the territory they served. The limits of a rainchief's authority seldom extended beyond the territory encompassed by several adjacent ridges, whereas later chiefs were able to establish much more extensive spheres of influence. These geographical limitations may well reflect the limited nature of their economic resources. The agricultural commodities that rainchiefs received for their services, while providing for a degree of economic differentiation and supporting redistributive and alliance-building activities, were, in general, too perishable, and too pervasive within the society at large, to serve as an effective political resource. At the same time, the localized nature of rainfall patterns, and particularly the distribution of thunderstorms within the Bashu region, may have caused the power of individual rainchiefs to be associated with specific local areas and thus to have set conceptual limits to the size of their chiefdoms.

Finally, the political and jural authority of rainchiefs was neither as extensive nor as institutionalized as that which was eventually achieved by later Bashu chiefs. Rainchiefs were first and foremost ritual leaders, whose influence in the daily lives of their clients is said to have been minimal.

Despite these differences, Bashu chiefship at the end of the eighteenth century was moving conceptually toward the type of chiefship that was to emerge during the nineteenth century and that was described in chapter one. The completion of this evolution was stimulated at the beginning of the century by the arrival of a second group of migrants, whose economy was initially based on pastoral activities, and by patterns of competition that their arrival initiated.

Like the Nande cultivators who had preceded them, these herdsmen expanded their settlements up the river valleys that tied the mountains to the plains. Unlike their predecessors, however, they possessed extensive economic resources, primarily in the form of livestock, which permitted them to compete successfully with the rainchiefs who controlled the region and to transform the nature and scale of political leadership. To understand this transformation and complete our examination of the historical background against which the politics of ritual chiefship was played out during the nineteenth and early twentieth centuries, we must now turn to the forces and events that surrounded the arrival of these new migrants, who were to become the present chiefs of the Bashu region.

CHAPTER THREE

THE POLITICAL TRANSFORMATION
OF THE UPPER SEMLIKI VALLEY REGION

The movement of herdsmen into the Mitumba Mountains and the subsequent transformation of the Bashu chiefdoms at the beginning of the nineteenth century followed changes in the economic and political environment of the upper Semliki Valley during the seventeenth and eighteenth centuries. As noted in chapter two, periods of drought, accompanied perhaps by outbreaks of sleeping sickness, during the seventeenth and early eighteenth centuries altered the ecological and political balance between herdsmen and farmers in the valley, causing many Nande/Konjo cultivators to move their settlements up onto the slopes of the mountains and northward into the forest. At the same time, other cultivators became the clients of herdsmen, whose economy came to dominate the valley. These changes culminated in the consolidation of Bahima-Babito authority over Busongora and the establishment of the Babito Kingdom of Kisaka at the beginning of the eighteenth century.

Pastoral domination of the valley was accelerated during the eighteenth century by the arrival of new groups of herdsmen fleeing from the political turmoil that disrupted the states of western Uganda during this period. The Nyoro invasion of Nkore during the reign of Ntare IV of Nkore (1713–40), Nkore's invasion of Buweju during the same period, the succession wars in Nkore that followed the death of Ntare IV, and the civil wars that followed the break-up of Mpororo at the end of the eighteenth century all contributed to the political instability of western Uganda at this time and stimulated the migration of many pastoral immigrants into the upper Semliki Valley.[1] The arrival of these new immigrants contributed to the social and economic transformation of the valley and to the further dominance of the herdsmen over Nande/Konjo cultivators. During the last years of the eighteenth century and the beginning of the nineteenth, this shift in the political balance between the valley's two major economic groups was completed by the consolidation of the herdsmen's political control over the west bank of the Semliki

72

Valley under the leadership of the Babito princes of Busongora and their Bamoli allies. Both of these ruling groups subsequently extended their political influence into the Mitumba Mountains.

BABITO EXPANSION
INTO THE UPPER SEMLIKI VALLEY

The expansion of Babito authority west of the Semliki River and the establishment of the Babito chiefdom of Bugaya over the northern half of the Semliki's west bank began during the reign of Mairanga, who ruled Kisaka at the beginning of the nineteenth century. Additional pressure for Babito expansion to the west of the Semliki occurred during the reign of Mairanga's successor, Kioma, as a result of military incursions into Busongora by the armies of Kaboyo, the newly crowned king of Toro. Kaboyo, who had successfully established his independence from his father, the king of Bunyoro, around 1830, wished to solidify his political position by acquiring control of the salt and cattle resources of Busongora (Kamuhangire, 1972: 11–12; Ingham, 1975: 28–30). His raids into the region during the 1830s and 1840s caused many Babito from Kisaka to seek refuge to the west of the Semliki, thus furthering the expansion of Babito authority over this region. Other Babito refugees fled northward along the western slopes of the Ruwenzoris.[2]

A second chiefdom, called Kiyanja, was established during this same period over the southern half of the Semliki's west bank by members of the Bamoli clan who were clients of the Babito princes of Kisaka. The alliance between the Bamoli and the Babito appears to be of considerable antiquity, perhaps dating from the establishment of Babito rule in Bunyoro. Large concentrations of Bamoli live in northern Bunyoro and identify with the *ngabi* (bushbuck) totem of the Babito. According to Buchanan, the Bamoli may have been some of the Babito's earliest allies in the region. The Bamoli were also associated with the establishment of Babito rule in the Kingdom of Bwera to the southeast of Bunyoro, although there is some disagreement as to the nature of this association. Nyoro sources indicate that when the Babito expanded their influence over Bwera they placed a Mumoli chief as governor of the region (Buchanan, 1974: 219). Bwera traditions, on the other hand, suggest that the Bamoli were already the dominant clan in Bwera and were recognized as such by the Babito. This claim is reinforced by the fact that the head of the Bamoli clan in Bwera was also the principal medium for the spirit of the Cwezi ruler Wamara, indicating a prior association between the

Bamoli and the pre-Babito Cwezi. Bamoli influence in Bwera may, there-fore, predate Babito expansion. In any case, the Bamoli in Bwera, like those in Bunyoro, were allied with the Babito, who frequently visited the head of the Bamoli clan at Masaka Hill in his capacity as Wamara's medium. The Nyoro rulers also sent representatives to Masaka to acquire talismans and herbs used in their accession ceremonies. Wamara's priest in turn obtained these objects from the islands of Lake George.[3]

A similar relationship developed between the Bamoli and the Babito princes of Busongora, though it is unclear whether the Bamoli preceded the Babito into the region. A Mumoli priest living on Kakule Island in Katwe Bay served as a medium for Wamara, the dominant deity in Busongora. The Babito of Busongora also sent representatives to Kakule Island to acquire objects for their accession ceremonies. Like Wamara's representative in Bwera, the priest of Kakule Island obtained these objects from the islands of Lake George and in particular Ntsinga Island. In return for these services, the Babito recognized Bamoli influence over the islands and lake shore around Katwe.[4]

The expansion of Bamoli settlements to the west of the Semliki River and the foundation of Kiyanja were apparently initiated by the arrival of a new group of migrants from the south end of Lake Edward during the middle years of the eighteenth century. This group, known locally as Bakingwe, were engaged in lakeside commerce and were related cul-turally to the Nande/Konjo peoples who had formerly occupied the Semliki Valley. The Bakingwe settled around Katwe under the leader-ship of members of the Barenge clan, who trace their origins to Rwanda and Ndorwa.

The Bakingwe took an active interest in the transport of salt and soon expanded the dimensions of the salt trade, using fast, lightweight canoes, which they are said to have introduced into the region to convey salt to markets that were established along the lake shore. These markets became centers for the exchange and distribution of salt throughout the hinter-land regions that extended beyond the lake shore.

The success of the Bakingwe in expanding the salt trade brought them wealth and prestige, which they used to extend their political influence over the Katwe region. While they evidently accepted the overlordship of the Babito, they acquired direct control over the islands and lake shore around Katwe on which the Bamoli lived, and according to Kamuhan-gire, became chiefs over the region.[5] The Bamoli retained substantial influence in the region because of their role as mediums for Wamara, but they were officially subordinated to the Bakingwe. This subordination apparently encouraged the Bamoli to expand their settlements to the west of the Semliki River, where they established the chiefdom of Kiyanja at the beginning of the nineteenth century.[6]

By the early years of the nineteenth century, therefore, the Babito and their Bamoli clients had established their authority over most of the upper Semliki Valley. According to Bashu traditions, the political expansion of these two groups did not stop here. Instead, both groups expanded their political influence into the mountains that bordered the plains and eventually established chiefdoms in the Mitumbas. The Bamoli are said to have expanded their political influence up the valley of the Tumbwe River into the southern part of the present Bashu region, establishing settlements at Kikotski, Kinyerere, and around Mount Kasongwere. They also settled further to the south along the lake shore, where they established chiefdoms at Vukendo and Vukununu. Toward the end of the nineteenth century a later Bamoli expansion occurred into the mountains south of the Karuruma River.[7]

The Babito are said to have established chiefdoms along the western slopes of the Ruwenzoris during the early decades of the nineteenth century. A junior branch of this Babito group, known locally as Banisanza, subsequently crossed the Semliki Valley and settled around Mount Lisasa, to the north of Isale, after attempting unsuccessfully to establish themselves in the plains just south of the equatorial forest. Two other groups that claim to have Babito connections expanded their settlements into the mountains from the plains chiefdom of Bugaya during the early years of the nineteenth century. The first of these, led by a man named Mukumbwa, settled south of Isale near Luvere in Maseki. The second, and more important, represented in Bashu traditions by a man named Kavango, settled at Kivika on the slopes of Mount Mughulungu in Isale. Kavango's descendants eventually expanded their political authority over most of the present Bashu region.[8]

While Bashu traditions are unanimous in tracing the origins of their present chiefs to the Babito and Bamoli groups who ruled the upper Semliki Valley at the beginning of the nineteenth century, they have to be used with care. To begin with, the traditions are consistent with Bashu ideas concerning east and west. The "east," and thus the direction of the Semliki Valley, is associated with ritual purity, birth, and renewal, while the "west" is associated with pollution, ritual danger, and death. The "east" is thus prescribed as the direction from which one's ancestors must have come.[9] In addition, there is the possibility that the claims of mountain rulers to being related to the chiefs of the plains represent an attempt by mountain chiefs to increase their prestige.

Fortunately, independent information comes from the traditions of the pastoral groups who formerly occupied the Semliki Valley. While these groups no longer inhabit the valley, having been forced to leave on account of sleeping sickness in 1932, their traditions were partially collected by Belgian agents at the end of the nineteenth and beginning of the twentieth century. In addition, I was able to collect a few traditions from former herdsmen who settled in the mountains following the evacuation.[10] According to Belgian reports, the Babito chief Kalongo, who was settled at Kivere in the chiefdom of Bugaya at the time of the initial Belgian occupation of the valley, claimed that he and the descendants of the mountain leader Kavango were members of the same family, Kavango's branch having settled in the mountains, leaving Kalongo's ancestors to rule the plains.[11] Similarly, a descendant of the plains chief Isimbwa, whose settlement was located in the southern part of Bugaya, supported the traditions of Mukumbwa's family, claiming that Isimbwa travelled with Mukumbwa from Kitara and that Mukumbwa settled in the mountains while Isimbwa remained in the plains.[12] The historical connection between Isimbwa and Mukumbwa is also supported by a Belgian report that states that during the early years of Belgian administration in the mountains, Mukumbwa's descendants avoided contact with Belgian administrators by returning to the plains to live with their 'Hima' relatives.[13] Finally, the more recent expansion of Bamoli authority into the mountains toward the end of the nineteenth century is supported by a Belgian report that notes that the leader of the group, Molekela, had political authority in both the plains and the mountains. "The chiefdom of Molekela is divided into two parts, one is in the plains, the other in the mountains. In the plains are the animals and the Wahima subjects who guard them. The mountain subjects are of the Wanande race. . . ."[14]

The connection between the present Bashu chiefdoms and Babito rule in the upper Semliki Valley is further supported by ethnographic data. While Bashu chiefship is culturally similar to chiefship found elsewhere in the lake region of Zaire, and especially among the southern Nande, Hunde, and Nyanga, there are several features of Bashu royal ritual and regalia that distinguish Bashu chiefs from their counterparts elsewhere in the region, and that indicate an historical connection with the Babito rulers of western Uganda. Among these elements are the use of the terms *mukama* to refer to a chief, *mubito* for the chief's eldest brother and ritual representative of the ruling lineage, and *kikali* for the chief's homestead.

These terms do not occur elsewhere in the lake region of Zaire, but do occur in western Uganda, where they are closely associated with Babito rule.

Babito influence among the Bashu is also indicated by the practice of removing and preserving the lower jawbone of a deceased *mwami*. While a number of Zaire lacustrine people remove the *mwami*'s whole skull, the removal of just the lower jawbone occurs primarily to the east, among the Babito-Banisanza of the Ruwenzoris and among the ruling families of Bunyoro, Busongora, Busoga, and Kiziba. The Baganda also preserve the king's lower jawbone, but, like the lacustrine Zaire states, remove the whole skull. The specific practice of removing the lower jawbone, therefore, appears to be an element of Babito royal culture.

Finally, the relationship between Kavango's descendants and the Babito of Bugaya and Kisaka is supported by the fact that Kavango's descendants have to acquire a piece of barkcloth and certain other objects used in the royal accession ceremonies from an *mbandwa* medium who lives at a place called Itsinga Island. The similarity between this Itsinga and the Ntsinga Island in Lake George—from which the Babito rulers of Kisaka and Bunyoro obtained objects for their own accession ceremonies— suggests that Itsinga and Ntsinga may be the same place. If this is true, it would support the Babito identity of Kavango's descendants. Bashu informants, however, equate Itsinga Island with Kakule Island in Lake Edward, dispelling the notion that Itsinga and Ntsinga are one and the same. This does not mean, however, that Kavango's descendants did not acquire objects from the islands of Lake George. For neither the Babito of Bunyoro nor those of Kisaka visited the islands of Lake George them- selves. Instead, they employed the Bamoli mediums for Wamara as inter- mediaries. The medium for the Kisaka region was, as noted above, located on Kakule Island. The Bashu may therefore have acquired ritual objects from Itsinga/Ntsinga Island on Lake George, only instead of going there themselves, they, like the Babito of Kisaka, visited the Bamoli medium *(mbandwa)* living on Kakule Island, who in turn visited Ntsinga. With time it is likely that Kakule Island became identified with Ntsinga within the context of Bashu royal ritual. It is, of course, possible that the practice of visiting Itsinga predates Babito settlement in the region, and thus that the practice does not indicate a Babito connection. However, none of the earlier lines of chiefs among the Bashu make any claim to having visited Itsinga as part of their own succession rites.[15]

Ethnographic data, therefore, appear to support the contention put forth in Bashu traditions that their present chiefs have a close historical connection with Babito rule in the Semliki Valley. It must be remembered, however, that there are gaps in the ethnographic record of this region, the filling in of which might provide a rather different picture of the region's cultural development. Ethnographic evidence must therefore be seen as suggestive rather than conclusive.[16]

Finally, neither the collection of traditions from the plains nor the ethnographic evidence proves that the chiefs who settled in the mountains at the beginning of the nineteenth century were actual members of the Bamoli and Babito clans. They may have been other herdsmen who moved out of the valley in order to lessen Babito/Bamoli domination, or perhaps, given the close relations that evidently existed between the herdsmen who moved into the mountains and their counterparts in the plains, clients of the Babito and Bamoli, whose settlement in the mountains was directed by these plains chiefs. Either way, they would have taken along certain elements of Babito political culture acquired through association with the Babito and Bamoli.

Despite these difficulties, I have chosen to refer to the descendants of Kavango and Mukumbwa as Babito, since both groups refer to themselves in this manner, i.e., Bashu-Babito. For the same reason I shall use the term Bamoli to refer to those mountain groups that claim to be descended from the Bamoli of the plains.

HERDSMEN TO CULTIVATORS: THE MAKING OF MOUNTAIN CHIEFS

Bashu oral traditions provide little in the way of direct information as to why Babito and Bamoli groups expanded their settlements and political influence into the mountains. There is, however, indirect evidence that allows us to reconstruct several possible explanations for this sequence of events.

While the Mitumba Mountains are generally unsuitable for cattle acclimated to the plains, the lower foothills and mountain valleys, in which the Bamoli and Babito first settled, provided good grazing lands up to an altitude of about 1500 m, once Nande cultivators had cleared away the forest cover. In addition, these highland pastures received more rainfall than the plains and thus provided good pasturage during the dry season months of December to February and May to August, when the grasslands below became parched and quickly exhausted. Consequently, plains herdsmen regularly moved their cattle and other livestock into these highland pastures during the dry season in order to maximize grazing resources and maintain their herds. With the coming of the rains they would move their herds back into the valley.[17]

Access to these pastures was limited, however, by the presence of Nande cultivators, who had preceded the herdsmen into the mountains and used the lower mountain areas to grow maize, eleusine, bananas, beans, and red sorghum. The interpenetration of herdsmen and cultivators was generally marked by the peaceful exchange of goods and services, the people of the mountains providing foodstuffs, beer, pots, and arrowheads in exchange for hides, milk, butter, and goats provided by the herdsmen.[18] The relationship between these economically diverse groups, however, was fraught with tension. Herdsmen often found themselves subject to mountain chiefs who, as prior occupants, claimed control over these ecologically liminal areas and demanded tribute payments from anyone using them. Second, the herdsmen's cattle occasionally destroyed the cultivators' crops. When this happened the herdsmen were subject to fines imposed by the mountain chiefs.[19] A third and more critical source of tension between the herdsmen and their mountain hosts was the tendency of young men from the mountains to become the clients of plains chiefs in order to acquire goats, which were used for bridewealth payments. The plains chiefs gave each client several cattle to look after. When a calf was born, the client was given a goat to keep as payment for his good service. This patron-client relationship was called *vuhereka*.[20] By attracting a clientele of mountain cultivators, the plains chiefs threatened the position of the mountain leaders to whom these cultivators had formerly paid allegiance, reducing both the mountain chiefs' tribute income and their body of supporters. In at least one case, described below, tensions arising from this loss of support resulted in warfare between a mountain chief and encroaching herdsmen. These tensions and the insecurity they caused may have encouraged the Babito and Bamoli chiefs residing in the plains to establish more permanent settlements in the highland pastures, under the control of a son or client, in order to protect their grazing rights.

On the other hand, the movement of herdsmen into the mountain pastures may have been encouraged by competition and conflict among various pastoral groups in the Semliki Valley. The migration of herdsmen into the valley continued into the nineteenth century as a result of increased political instability in western Uganda. Continued strife in the successor states of Mpororo, succession wars in Bunyoro and Toro,[21] the previously mentioned Toro secession war, and a long series of military expeditions into Busongora by her neighbors all contributed to the movement of pastoral groups out of southwest Uganda and into the Semliki Valley.[22]

There is no way of knowing how many herdsmen and cattle moved into the valley at this time, for in the last two decades of the century warfare between the armies of Bunyoro, Toro, the Rwandan chief Karakwenzi, and, to a lesser extent, Manyema ivory traders greatly reduced the area's livestock population.[23] Rinderpest may have also reduced the valley's cattle population, though its effects there appear to have been less drastic than elsewhere in East Africa. Thus the Belgian estimate that 1,500 to 2,000 head of cattle existed in the upper Semliki Valley in the 1920s[24] most probably represents a small fraction of the valley's population during the middle years of the nineteenth century. At this time the region was described in Toro and Busongora songs as "the land where cattle were milked while roaming, and where there were more cattle than could be milked; where ghee was used to make watering pans and people bathed in milk" (Kamuhangire, 1972: 6). While this is an idealized view of "Eden," Casati is reported to have told Stanley that he once saw a Nyoro raiding party return from Busongora with many thousands of cattle (Stanley, 1890: 338).

If the cattle population of the valley was markedly larger during the early and middle years of the nineteenth century than it was when the Belgians made their estimate, there may have been competition for grazing and water rights, which would have increased during the periods of extended dry weather that were evidently common throughout East Africa during the early years of the nineteenth century (Nicholson, 1976: 96; Herring, 1979: 59–60). Within the context of this competition some groups may have chosen to acquire control of the highland pastures that border the valley as a means of escaping domination by stronger groups, or to mitigate the effects of drought on their herds, or perhaps even because the highlands provided more defensible locations for their settlements.

Whatever their motivation may have been, the groups who settled in these highland pastures continued their herding activities. Moreover, it is clear from the few marriage alliances that can be reconstructed for these herdsmen that their social and political orientation remained fixed on the world of the plains, Most of these marriages were with other plains groups and not with mountain cultivators. For example, Kavango's successor, Mukunyu, chose both the biological and ritual mother *(mombo)* of his successor from plains groups despite his settlement in the mountains.[25]

Nonetheless, the herdsmen eventually found it necessary to use their wealth in livestock to create additional alliances with influential mountain leaders and by so doing secure their position in the mountains.[26] In this way the herdsmen gradually extended their influence into the mountains and became increasingly involved in mountain politics. This process can be seen in the history of Bamoli expansion into the southern Bashu region under the leadership of a chief named Kadsoba at the end of the nineteenth century and will be examined in greater detail in chapter four.

Kadsoba's family settled in the plains and foothills that bordered the territory of the mountain chief Binga sometime between 1870 and 1880. Kadsoba's wealth in livestock attracted a large number of Binga's followers, mostly young men, who settled in the foothills and plains in order to become his clients and thereby acquire goats for bridewealth payments. While this arrangement proved satisfactory for Kadsoba and his mountain clients, it posed a threat to Binga, whose political power vis-à-vis other mountain chiefs was weakened by the movement of his subjects to the plains. Binga therefore attacked Kadsoba's settlements. Kadsoba, either unable to respond to these attacks militarily, or perhaps concerned that a protracted struggle would endanger his herds, decided to placate Binga by sending him gifts of cattle. Binga accepted these gifts and ceased his raids. However, Kadsoba did not wish to see his herds reduced through continued tribute payments, nor did he wish to abandon his settlement in the dry season pastures. He therefore established a series of alliances with other mountain leaders who were Binga's supporters. Through these alliances Kadsoba slowly undermined Binga's chiefship. Binga's death soon after his initial attacks on Kadsoba's settlement, and the succession of a young and inexperienced son to his position, no doubt aided Kadsoba's incursions. By 1900 Kadsoba's influence extended into the mountains and his followers had begun to establish themselves as mountain chiefs.[27]

While the desire to secure dry season pastures or to escape competition and drought conditions in the plains may have initially drawn pastoral groups into the mountains, these factors are unlikely to have stimulated their continued expansion into the upper reaches of the mountains, since these areas were not well suited to pastoral activities. Nevertheless, it is clear from Bashu traditions and from the location of royal burial sites that with each succeeding generation, the former herdsmen moved higher and higher into the mountains. Thus, while Kavango was buried at Kivika near the plains, his successor, Mukunyu, was buried at Kavale, and Mukunyu's son Luvango at Kalambi, each place higher than the preceding one.

Similarly, while Mukumbwa was buried near the plains at Luvere, his son Mandwa was buried at Muyina, which is located above 2000 m.

The continued expansion of former herdsmen into the mountains may indicate that pacifying the hinterland in order to secure dry season pastures proved to be a never-ending task. Having neutralized the power of one mountain chief, the Babito and Bamoli became a threat to other chiefs further up in the mountains. They may thus have found it necessary to expand the limits of their influence still further and become more involved in mountain politics. On the other hand, as we shall see in the history of Kavango's family, the descendants of former plains chiefs may have become involved in mountain politics as a means of establishing an independent support base with which to compete with rival siblings. The attraction of a mountain power base was particularly strong for sons who failed to inherit authority within the context of plains politics.

Whatever the reason for this secondary expansion might have been, it eventually led to a division between those descendants who remained in the plains and foothills and those who settled higher in the mountains. Those who became chiefs in the upper agricultural areas adopted many of the political institutions and values of the agricultural world in which they settled in order to consolidate and legitimize their control over this world. Consequently, their political institutions and practices more closely resemble those of their Nande neighbors in the mountains to the south and the west, and to a certain extent those of the Nyanga and Hunde, than those of their pastoral relatives who remained below and in Busongora. This divergence in political development ultimately led to the emergence of two chiefly lines within each family, one associated with the mountains and agriculture, the other associated with the plains and cattle keeping.

CONCLUSION

The early years of the nineteenth century saw the gradual expansion of Babito and Bamoli political authority into the Mitumba Mountains and the transformation of former herdsmen into mountain chiefs. In the years that followed, one group of Babito, descended from the former plains chief Kavango, succeeded in employing their access to livestock and other resources, and their consequent ability to form alliances with important mountain leaders, to establish control over Isale and surrounding regions. In doing so, they transformed the Bashu political system, increasing the geographical scale of political leadership and altering Bashu conceptions about the nature of chiefship.

PART THREE

THE POLITICS
OF RITUAL CHIEFSHIP IN ISALE

CHAPTER FOUR

RITUAL CHIEFSHIP AND THE POLITICS OF DOMINATION IN ISALE

Recent studies of state formation in precolonial eastern Africa have illustrated the importance of alliances as instruments of political expansion (Karugire, 1972; Feierman, 1972, 1974; Cohen, 1977). In so doing, they have successfully modified earlier studies that stressed the role of military force in the creation of states (Oliver, 1963; Kagame, 1963; Ford and Hall, 1947). Yet our knowledge of the actual process of alliance formation has been limited by an absence of extensive data on the composition of alliance networks. The early date at which many states were formed has made it difficult to obtain more than a partial list of some of the groups involved in the creation of particular networks. More complete lists, and data on the importance of these groups, are generally unavailable.[1] The absence of more detailed information has prevented historians from determining why particular groups were chosen as allies, what criteria determined the social and geographical distribution of alliances, and what, from an indigenous viewpoint, was the purpose of alliances. In short, the absence of data has prevented the construction of a clear picture of the motives and organizational principles or logic behind the formation of alliance networks. This in turn has contributed to the acceptance, often unconscious, of a formalist view of alliance formation, in which alliances among the Baganda, Asante, or Swazi, for example, are seen to be like alliances in modern Britain or France, in that they are simply created to build support and to undermine the support of rivals. At a certain level of abstraction this is undoubtedly true. However, I suggest that a closer examination of specific alliance networks reveals important cultural differences in the ways in which they are constructed and the purposes they serve.

In examining the history of Babito political expansion in the Mitumba Mountains under the leadership of Kavango and his descendants during the early and middle years of the nineteenth century, I have attempted to define in precise terms the historical status of the various groups involved in the alliance networks created by Babito chiefs. This type of reconstruction has been possible because of the relatively late date of

Babito expansion into the Mitumbas, and because, as noted in the introduction, there is a tendency to preserve the historical identity and importance of former allies in Bashu oral traditions and in the ritual structures of chiefship.

The resulting picture suggests that while the creation of alliance networks ultimately resulted in the acquisition of political power, their direct purpose, once the Babito became actively involved in mountain politics, was the creation of spheres of ritual control over the land. While alliances were formed with a variety of groups and individuals, the core of each network was composed of groups that possessed ritual influence over the land: clearers of the forest, diviners and healers of the land, and rainmakers. By creating alliances with these ritually influential groups, the Babito were able to coordinate ritual control over the land and in effect "cover the land," which, from a Bashu viewpoint, insured its well-being. This in turn allowed the Babito to acquire and maintain popular support. For while the creation of these alliances did not, from an external viewpoint, prevent the onset of famine, it did, as we shall see in chapters five and seven, provide successful chiefs with ritual credibility, which allowed them to deflect responsibility for disasters and thus maintain their legitimacy. Bashu ideas about the ritual role of chiefship, therefore, appear to have affected the choice of allies and the nature of alliance networks. Moreover, as I will suggest at the end of the chapter, they may also have shaped the geographical distribution of alliances.

THE FOUNDATION OF BABITO RULE

The success of Kavango's family in establishing their political control in Isale took several generations to achieve. Kavango and his immediate successor did little to establish wide spheres of ritual influence, and thus political control, in the mountains, their social and political activities being oriented primarily toward the plains. It was Kavango's grandson Luvango who first established the foundation for wider ritual and political control in Isale. Yet even this foundation, extensive though it was, was ephemeral, for it was based primarily on Luvango's personal ties with local ritual leaders and lineage heads and depended upon his presence. When he died, succession struggles split this foundation, and it was not until the turn of the twentieth century that the ritual basis of Babito chiefship was solidified, expanded, and institutionalized under the leadership of Luvango's grandson Vyogho. Nonetheless, achieving and maintaining this wider sphere of ritual influence was a primary political concern that helped shape political relations among Kavango's descendants as well as between the Babito and other Bashu chiefly lines during most of the nineteenth and early twentieth centuries.

The political environment of Isale at the time of Kavango's arrival was dominated by several rainchiefs. Each of these chiefs had among his supporters a number of lesser political and ritual figures. The dominant political figure in the region around Mount Mughulungu, where Kavango's family first settled, was the rainchief Mukirivuli. Mukirivuli's supporters included the family of Kaherataba, who were 'clearers of the forest' in the Mughulungu area and thus had direct ritual control over the land on which Kavango settled; the Baswaga of Katikale, who were also 'clearers of the forest'; the Bito of Biabwe, whose head had the power to divine the causes of crop failure and possessed ritual seeds *(mikene)* used in annual planting ceremonies; and the Bashu of Manighi, who were 'clearers of the forest' in Manighi and Biabwe. The Bashu of Manighi were also ritual specialists for Mukirivuli's family, guarding their royal regalia and directing their funeral and accession rites. Each of these groups depended on Mukirivuli's rainmaking skills and looked to his leadership in the coordination of land rituals.[2] This, then, was the political and ritual geography of the region into which Kavango moved his herds during the early years of the nineteenth century.

While Kavango's wealth in livestock, repeatedly mentioned in Bashu traditions,[3] provided the economic base for his family's expansion into the mountains, he may also have possessed two other political resources: rainmagic and his apparent relationship with the Babito rulers of Busongora and Bugaya. Kavango's descendants claim that their ancestors were rainmakers in the plains and that Kavango brought rainmagic with him when he settled in Isale. This claim is challenged, however, by Mukirivuli's descendants, who claim to have provided Kavango's family with rainstones,[4] and by a Muhima informant whose ancestors formerly herded cattle in the plains and who stated that the people of the plains always consulted mountain rainmakers during periods of drought or excessive rain.[5] While these claims call into question that of Kavango's descendants, they do not eliminate the possibility that Kavango possessed some form of rainmagic when he arrived in the mountains. To begin with, Stuhlman (1894: 282) reported finding rainstones in use in the plains. In addition, some of Kavango's descendants are said to have made rain with cattle horns and rainspears, techniques that were used by certain pastoral groups in western Uganda.[6] Kavango's family may have brought these techniques with them and subsequently acquired rainstones from

Mukirivuli's family. Finally, the statement of the Muhima informant that people of the plains depended on mountain rainmakers does not necessarily mean that they did not have rainmakers of their own. It may simply reflect the association that exists between the mountains and the rain. Thus, when asked why his ancestors depended on mountain rainmakers, the informant answered that this was because "rain comes from the mountains."[7] This answer reflects the impression one has in the plains that rain moves west from off the slopes of the Ruwenzoris. In addition, the heights of both the Ruwenzoris and Mitumbas are covered with clouds for a good part of the year and receive more rain than the plains. These climatic patterns may have made the power of mountain rainmakers appear greater than that of plains rainmakers, which in turn would explain why the people of the plains consulted mountain rainmakers when the rains were late in starting or ending, even though they had their own rainmakers in the plains. It is, therefore, possible that Kavango possessed some rainmaking skills before he entered the mountains. These skills may have helped him and his descendants establish their authority in the mountains by providing them with ritual legitimacy.

Kavango's apparent association with the Babito rulers of Bugaya and Busongora may have been useful in two ways. First, it provided him with prestige. Thus, it is often said that the people of Isale accepted the political authority of Kavango's family because they had been chiefs in Kitara, a name that is used to describe the land beyond the Semliki, and particularly Busongora and Toro. Secondly, and perhaps more significantly, Kavango's connection to the Babito appears to have allowed his family to provide their supporters with superior access to the salt lake at Katwe, and thus to salt, which was an important trade item among the Bashu, as it was throughout the western Lakes Plateau Region. To understand the nature and importance of this salt connection, it is necessary to describe briefly how the salt trade operated.

The salt trade between Isale and Katwe was open to anyone who possessed the energy and desire to travel to Katwe, and thus, unlike the lakeside salt trade, was not controlled by a specialized trading group. It was not even necessary to have goods to trade for the salt since it was possible to mine salt for oneself and then provide a portion of the extracted salt to the chiefs of Busongora.[8]

The men who travelled to Katwe usually went during the dry season because of the lull in agricultural activities during this period and because the best quality salt was obtainable during the dry season. Travel up and down the mountains was also easier at this time. The traders usually travelled in large groups armed with spears for protection against wild animals and occasional attacks by the inhabitants of the plains. By the end of the nineteenth century, by which time the valley had become a battleground for competing political and commercial groups, the number of traders in a salt caravan reached fifty or more.

Arriving at Katwe, the traders who had brought goods—primarily agricultural produce, such as beans, sorghum, and eleusine—traded their goods with the people who lived permanently at Katwe in return for salt. As Stanley noted in 1889, no crops were grown in the area immediately surrounding the lake, so that the people of Katwe were dependent on trade with neighboring mountain peoples for food (1890: 339–40). The men who had nothing to trade arranged with the guardians of the lake to extract their own salt. Whether one extracted salt or traded for it, the transactions and activities took several days to complete. During this time the mountain traders camped at sites assigned by the market master, *muhoza*, who oversaw all commerce at Katwe for the Babito rulers of Kisaka.

Once all the members of the trading party had completed their transactions, the caravan reassembled and recrossed the Semliki Valley. Returning home, a trader would distribute some of his salt to the members of his family and send a gift to the head of the group from whom he had received land, as well as to his chief. The rest of the salt would be used for his own consumption and for trading for locally produced specialty goods, such as pots, baskets, iron tools, and even beer.

While the salt trade was not a specialized activity, some Bashu men participated in it to a greater extent than others. More active traders travelled more frequently to Katwe and took the salt they acquired to more distant regions of the mountains to trade for goods not available in Isale. Some went west to the Ituri Forest in order to acquire resin oil, red powder, and fiber rings known as *vutegha*, used by Bashu women to decorate their arms and legs, and by all Bashu as a form of currency.[9] Other salt traders travelled south to Buhimba, where they obtained iron hoes and spearheads from the highly skilled smiths of this region.

Whether a man engaged in salt trading on a regular or an occasional basis, it provided a way of acquiring important nonsubsistence goods. Moreover, because salt could be traded for goats, participation in the salt trade was a means of acquiring bridewealth and thus a road to social advancement. The absence of any initial capital expenditure requirement meant that it was a road that was especially attractive to young men who had not had time to acquire other forms of wealth. Participation in the salt trade was thus an important aspect of Bashu economic and social life.

Kavango's family, through their connection with the Babito rulers of Busongora and Bugaya, were evidently able to improve their supporters' access to the salt trade in two ways. First, they could help guarantee safe passage through the plains, which were controlled by the Babito. Second, they could evidently decrease the cost of obtaining salt. According to Kamuhangire (1975: 76), every trader who came to Katwe had to pay a tax to the market master. If a person chose to mine salt, he paid one-third of his accumulated salt to the market master. If he brought his own goods to trade, a third of the goods would go to the market master. Bashu informants, however, are unanimous in stating that they were exempt from the tax because the salt lake was divided between the rulers of Busongora and their own chiefs, meaning Kavango's descendants. Only after the British occupied Katwe at the end of the nineteenth century were the Bashu required to pay a tax on their trading or extracting activities. The existence of this exemption indicates that Kavango's family, through their Babito connection, were able to offer their supporters cheaper salt as well as safe passage to and from the lake, and thus possessed an important resource with which they could attract supporters and expand their political influence in the mountains.[10]

Despite these varied resources, Kavango did little more than secure his family's access to the lower pastures of Mount Mughulungu. While it is possible that he was unable to overcome the resistance of mountain chiefs, his expansion may have also been restricted by the primacy of his pastoral interests.

Kavango initially acknowledged the authority of existing clan leaders and chiefs in return for the right to graze his herds in the highland pastures around Mughulungu. Thus he paid the *muhako*, a payment of several goats given annually to 'clearers of the forest,' to Kaherataba, the *mukonde* of Mughulungu. A similar payment was made to the rainchief Mukirivuli as *ngemu*. At the same time, Kavango strengthened his position in the highland pastures and secured future access to them by establishing relationships with three other mountain groups.

The first of these groups was the Batangi of Lulinda, a village located on the southwest slopes of Mount Mughulungu. The origins of this alliance evidently preceded Kavango's arrival in Isale, for the Batangi claim to have travelled with Kavango's family from Busongora, where they had been *bakaka* (royal buriers) and directed the performance of royal rituals for Kavango's ancestors.[11] This claim is supported by the fact that the Batangi are also *bakaka* for the Babito chiefs of the western Ruwenzoris[12] and by the fact that the *mombo*, ritual wife, of Mukunyu

(Kavango's successor) came from the Batangi of Busongora.[13] On the other hand, there were evidently other Batangi already settled at Lulinda, for Kaherataba's descendants claim that their ancestors intermarried with the Batangi of Lulinda before Kavango arrived in Isale, and that both Kaherataba and his father, Kabahetwire, took their first wives from this group.[14]

It thus appears that Kavango used his connection with the Batangi of Busongora to ally himself with the Batangi of Lulinda, who, as Kaherataba's maternal uncles, may have assisted Kavango in gaining access to the land that Kaherataba controlled. This alliance was cemented by a marriage between Kavango and a woman from Lulinda, who bore Kavango's successor, Mukunyu.[15]

Kavango further strengthened his position in the mountains by becoming the client of a second rainchief, Muhiyi (also called Mutsawerya), who was Mukirivuli's major rival in the Mughulungu area. Kavango acknowledged Muhiyi's ritual authority by paying him annual tribute *(ngemu)* and participating in the land rituals he directed. By acknowledging the authority of both Mukirivuli and Muhiyi, Kavango insured himself against the attacks of either one. From the perspective of Mukirivuli and Muhiyi, Kavango's dual allegiance may have been distasteful. However, it is likely that neither side wished to give up its tie with this wealthy client, whose access to livestock and salt could be used to augment their own position in the mountains.

The importance of Kavango's alliance with Muhiyi is reflected in the traditions that describe the founding of the present Bashu chiefdoms by Kavango and his descendants. As noted in the introduction, these traditions describe Muhiyi as Kavango's brother or son and give Muhiyi a prominent role in the expansion of Babito rule into Isale. The traditions, in fact, center on Muhiyi's activities, though they end with the establishment of Kavango's authority. Since there is little in the subsequent history of relations between Muhiyi and Kavango's descendants to indicate that Muhiyi and Kavango were actual kinsmen or that Muhiyi was a Mubito, these traditions probably attest to the important role played by Muhiyi's family in the establishment of Babito rule.

Kavango evidently formed a third alliance with a group of Bashu who claim to be descended from the original forest dwellers of the Mitumba Mountains, the Basumba. This group, who presently live at Makungwe on the slopes of Mount Mughulungu, claim to have welcomed Kavango when he arrived and to have given him fire. It is unclear, however, whether this last claim reflects historical fact or simply the role this family plays in the investitures of Kavango's descendants. It is the Basumba of Makungwe who relight the royal fires following the investitures of Kavango's descendants in the Mughulungu region and thus reenact the gift of fire. The Basumba in general are thought to have a close association with the land and especially the forces of the bush. This association accounts for their role in killing the ritually dangerous flying squirrel from which the *mbita* is fabricated.[16]

The alliances and associations that Kavango formed in the mountains secured his position in the foothills around Mount Mughulungu. He did not, however, attempt to establish direct control over these areas and was not invested as a mountain chief. He is, in fact, remembered primarily as a herdsman. Nonetheless, his alliances paved the way for future Babito expansion.

Kavango's successor, Mukunyu, was in a sense a transitional figure in the transformation of Kavango's family from herdsmen to mountain chiefs. On the one hand, his alliances and retention of certain ritual practices associated with the plains indicate that, like his father, he maintained his ties with the pastoral world of the Semliki Valley. On the other hand, he was also invested in the manner of mountain chiefs and adopted some of the political culture of the mountains.

Mukunyu formed marriage alliances with at least two mountain groups, the Baswaga of Katikale, who, as noted above, were 'clearers of the forest' to the west of Mughulungu, and the Bashu of Musitu in Malio. However, his most important allies were pastoral groups. Thus, he took his ritual wife, *mombo*, from the Batangi of Mbulio in Busongora, and chose as his successor the son of a wife from the pastoral Banisanza of Lisasa.[17] Had he been interested in expanding his influence into the mountains, he would presumably have chosen his *mombo*, as well as the mother of his successor, from one of the mountain groups with which he allied himself. Instead, he chose to use these important positions to strengthen his ties to the world of the plains. He further reinforced these ties by retaining the Batangi, who had accompanied his father from Busongora, as his royal ritual specialists and guardians of his regalia. Mukunyu is also said to have visited the medium for Wamara on Itsinga Island before his investiture, thus continuing the traditions of his pastoral ancestors, or former patrons, in Busongora.[18]

At the same time, however, Mukunyu increased his family's association with the political world of the mountains by being invested as a mountain chief. This was evidently necessary in order to establish the legitimacy of his claim to chiefship among the mountain cultivators. Mukunyu received the emblems of authority associated with mountain chiefship, including his *mbita,* from Mukirivuli. This is in keeping with Bashu ideas about the proper acquisition of authority, which decree that emblems of authority have to be obtained from a legitimate source, i.e., they have to be given by an invested *mwami* in order to be legitimate. A common way of attacking a *mwami's* legitimacy is to say that his *mbita* is not real (i.e., legitimate) because "he bought it," meaning that it was not given to him.[19]

In agreeing to provide Mukunyu with these emblems, Mukirivuli probably weighed a number of factors. First, in favor of his participation, was the fact that, in principle, givers of emblems of authority are politically superior to the receivers of those emblems (although in reality this superiority may be purely ceremonial if the givers of the emblems are not politically dominant in other, more practical, ways). By providing Mukunyu with the necessary emblems of chiefship, Mukirivuli could increase Mukunyu's political dependence on him and strengthen the patron-client relationship that existed between them. Moreover, having an invested chief as a supporter increased Mukirivuli's own ritual authority. On the negative side, Mukirivuli may have realized that by investing Mukunyu he was giving him legitimacy that he would not otherwise have and thus was making him a potential rival for political power. While we cannot be certain about what prompted Mukirivuli to participate in Mukunyu's investiture and to discount the potential threat that this decision created, he may have felt that if he did not consent to Mukunyu's request, Mukunyu would seek the assistance of Muhiyi's family, in which case he would acquire legitimacy without at the same time increasing his dependence on Mukirivuli. If this happened, Mukunyu would have been an even greater threat. Moreover, his investiture would reinforce Mukunyu's ties with Muhiyi and strengthen Muhiyi's position vis-à-vis Mukirivuli.

Along with the emblems of mountain chiefship that Mukunyu received at his investiture, he borrowed the practice of choosing a *mombo* to be the social mother of his successor. Mukunyu was the first member of his family to adopt this institution. By doing so, he altered the succession pattern, which up until this time is said to have followed the rule of primogeniture.[20] Borrowing this institution of mountain chiefship increased the legitimacy of Mukunyu's authority in Bashu eyes and moved him closer to the political culture of the mountains, even though the initial occupant of the office was from the plains. It also opened up the succession process, since, as noted in chapter one, the *mombo* was only the social mother of the *mwami's* eventual successor, who could be chosen from among the sons of any of the *mwami's* wives. Consequently, any allied group could acquire the politically influential position of maternal uncles to the next chief. This gave each ally a vested interest in the chiefship. Adopting the institution of the *mombo*, therefore, not only conformed to mountain practices but also improved the ability of Mukunyu and his descendants to forge and maintain future alliances with mountain groups and paved the way for greater involvement in the world of mountain politics. There was, of course, a potential cost to this practice. Opening up the succession process increased the possibilities for succession disputes. It is interesting to speculate whether the ideological view of the *mombo* as both preserving and weakening *bwami* during the interregnum period is related to the conflicting nature of her actual impact on the political strength of chiefship.

By increasing his associations with the world of mountain politics while at the same time maintaining one foot firmly in the world of the plains, Mukunyu can be seen as a transitional figure between his father —who was closely associated with the plains—and his son Luvango, who, as we shall see, completed his family's transformation from herdsmen to mountain chiefs.

THE EXPANSION OF BABITO RULE

Mukunyu had three sons, or at least three who are presently remembered among the Bashu. His firstborn son, Luvango, was the product of Mukunyu's alliance with the Baswaga of Katikale. The second, Kamesi, was the son of Mukunyu's wife from the Bashu of Musitu, and the third, Visalu, was born of his marriage alliance with the pastoral Banisanza of Lisasa. As potential rivals for the chiefship, each son was raised among his maternal uncles in order to protect him from possible attacks by his siblings' supporters. This practice also served to strengthen Mukunyu's alliances with his affinal relations and to extend his influence into the regions in which they lived.

The separation of Mukunyu's sons may have affected their subsequent careers. Among the Banisanza, Visalu served as a herd boy; when he grew older he was put in charge of a section of his father's herd, which he later inherited. Visalu thus learned the pastoral ways of his ancestors and, like his ancestors, showed little interest in expanding his authority beyond the foothill regions of the Mitumbas when he became chief.[21] Kamesi, on the other hand, was raised in Musitu, which was evidently a forest region of recent Bashu penetration. Thus the region was still dominated by forest Bantu peoples (Basumba) and Kamesi's political aspirations, whatever they may have been, were consequently restricted. In fact, Kamesi's descendants did not succeed in establishing widespread authority in the Malio region until the reign of Kamesi's grandson, Mbonzo, by which time most of eastern Malio had been cleared of forest. Even then, however, the descendants of former Basumba chiefs retained considerable authority over the region.[22]

Finally, Luvango was raised at Katikale, located in the heart of the Bashu political world. Here, in contrast to Visalu, Luvango was able to observe and be educated in the world of mountain politics. Moreover, in contrast to Kamesi, Luvango's political aspirations were not retarded by the dominant presence of a culturally distinct population. Luvango thus had a better opportunity than his brothers to establish himself as an important political figure in the mountains. He in fact grew to be a master at mountain politics, who clearly understood the ritual basis of political authority in Isale.[23]

Luvango's decision to construct a power base in the mountains was no doubt encouraged by the choice of Visalu as Mukunyu's successor. This choice reflected Mukunyu's continued orientation toward the world of the plains. However, it also represented a break with the pattern of primogeniture and was a product of Mukunyu's adoption of the office of *mombo*. Deprived of what he may have perceived as his rightful patrimony, Luvango chose to establish a power base in the mountains.

Luvango began his campaign with three initial resources. First, he had the Baswaga of Katikale for maternal uncles and thus the support of a ritually influential family of 'clearers of the forest.' Second, he was appointed *mukulu*, responsible for performing sacrifices for the *mwami* and representative of the royal lineage. This position gave Luvango considerable influence within his brother's chiefdom. Finally, Luvango was appointed temporary regent or guardian for Visalu, who is said to have been too young to exercise chiefly authority on his own at the time of Mukunyu's death. As both regent and *mukulu*, Luvango was in a position to employ his father's wealth in livestock and ritual status for his own benefit.[24]

Working with these initial resources, Luvango proceeded to construct an extensive network of alliances with lineage heads and ritual leaders throughout Isale. These alliances eventually tied together the northern and southern halves of Isale into a single sphere of ritual influence, which exceeded that of any existing chief and, from a Bashu perspective, made Luvango the dominant chief in the region. This success allowed Luvango to end his family's subordination to the descendants of the rainchief Mukirivuli and to usurp Visalu's claim to Mukunyu's chiefship.

One of Luvango's first alliances was evidently with the Bito of Biabwe, for it was this marriage that produced his eldest son, Mutsora. The support of this family was important for two reasons. First, as 'diviners of the land,' to whom the people of northern Isale went to determine the causes of crop failure, they could, as will be seen in chapter five, help a chief maintain his ritual credibility during a natural disaster. Second, the Bito of Biabwe were also 'healers of the land,' who possessed ritual seeds thought to insure the productivity of the gardens in which they were placed. The Bito of Biabwe were thus a ritually important group in northern Isale.[25]

Luvango formed a second marriage alliance with the Baswaga chiefs of Ngulo.[26] This alliance evolved out of his relationship to the Baswaga of Katikale and was important in two respects. First, it provided Luvango with the prestige and support that came from marrying the daughter of an established *mwami*. Second, it increased his political legitimacy within Isale, for the chiefs of Ngulo controlled the famous water oracle at Musalala, to which the people of Isale often went to obtain answers to difficult questions. When Luvango decided to take an *mbita* and declare himself to be the rightful successor to Mukunyu, the oracle at Musalala is said to have confirmed his claim, thereby helping to establish his legitimacy.[27]

A third alliance was formed with the Bahera of Ngukwe, from whom the mother of his chosen successor, Kivoto, came.[28] This alliance was important because the Bahera of Ngukwe were the senior branch of the very large Bahera clan, whose settlements dominated the southern half of Isale. By gaining the support of this group, Luvango laid the groundwork for the southern expansion of his political influence. This alliance may also have led to a later marriage alliance with the Bahera of Maseki, who were important allies and royal ritual specialists for the family of the Babito chief Mukumbwa. Luvango's son by this marriage, Kahese, grew up in Maseki and eventually succeeded in usurping the authority of Mukumbwa's family and establishing himself and his descendants as chiefs in the region.[29]

Luvango evidently established a fifth alliance with the Bashu of Mbulamasi, also located in southern Isale, for it was this group that provided Luvango with his *mombo*. They also presided over his funeral rites and became the guardians of his jawbone and other regalia following his death. The allocation of these important ritual roles to this group reflects their importance in Isale. Like the Bito of Biabwe, the Bashu of Mbulamasi were 'diviners of the land.' However, their influence extended over the southern half of Isale, whereas the influence of the Bito of Biabwe was in the northern half. A section of this lineage, moreover, had a reputation as rainmakers in the area around Mbulamasi.[30]

Luvango formed at least three other important alliances. The first was with the Bashu of Manighi, who claim to have provided Luvango with a wife and to have served as *bakaka* at Luvango's funeral. The Bashu of Manighi, who were 'clearers of the land,' had previously directed the performance of royal rituals for Mukirivuli's family, to whom Luvango's family still owed allegiance. This alliance, therefore, served two purposes: it undermined the support base of Luvango's overlord while increasing Luvango's legitimacy in the mountains. The alliance also benefited the Bashu of Manighi, for they continued to serve as royal ritualists for Mukirivuli's descendants and thus were able to maintain a strong position vis-à-vis both families, limiting their dependence on either one and playing one side against the other. It is evidently for this reason that they have never paid tribute to Luvango's descendants and, unlike other groups in Isale, have never been forced to accept one of Luvango's descendants as a subchief.[31]

The second alliance was with the Bashu of Mwenye. When they were involved in a war with the Basumba, who lived in the forest to the northwest of Mwenye, Luvango lent them a number of supporters, under the leadership of a client named Kanyabwarara of the Bahombo clan. After defeating the Basumba, Kanyabwarara settled in Mwenye and established himself as Luvango's representative in the region.[32]

The final alliance for which I could collect data was with the Banisanza of Lisasa, from whom Luvango also took a wife. This alliance followed a war with the Banisanza, which will be discussed below.[33]

Luvango's alliances provided him with a wide base of support and allowed him to end his family's subordination to the family of the rainchief Mukirivuli. To this end, his alliances with Mukirivuli's allies, the Baswaga of Katikale, the Bito of Biabwe, and the Bashu of Manighi, were particularly important, for they undermined the authority of Mukirivuli's descendants by creating conflicting loyalties among these groups.

Luvango's ultimate success in ending his political dependence on Mukirivuli's family, however, resulted from a succession struggle among Mukirivuli's sons, which weakened their position in Isale while strengthening that of Luvango. Mukirivuli was initially succeeded by his firstborn son, Kisere. However, Kisere's younger brother, Mbopi, disputed this succession and claimed to be Mukirivuli's legitimate successor. Luvango took advantage of this dispute by allying himself with Mbopi and agreeing to support Mbopi's claim to chiefship in return for Mbopi's participation in his own rather than Visalu's investiture. Acquiring the support of this section of Mukirivuli's family helped Luvango establish his claim to being Mukunyu's legitimate successor, for Mukirivuli had invested Mukunyu. At the same time, it increased the division between Mbopi and Kisere and thus weakened the authority of both men.[34] Finally, it added the ritual authority of this family of important rainchiefs to Luvango's growing sphere of ritual control in Isale.

Having weakened the authority of Mukirivuli's descendants and strengthened the position of his own family, Luvango moved against his younger brother Visalu, declaring himself to be Mukunyu's rightful successor. Visalu refused to give way to his brother's usurpation, and an armed struggle between Visalu's supporters and those of Luvango ensued. This conflict was won by Luvango, who forced Visalu to flee Isale and seek refuge among his maternal relatives at Lisasa. While the people of Lisasa rallied to Visalu's cause and launched a counterattack against Luvango and his supporters, the attack proved unsuccessful, and a settlement was eventually reached between Luvango and Visalu's maternal relatives. This settlement resulted in Luvango's marrying a woman from Lisasa, and in Visalu's continued exile.[35]

Luvango's use of armed force against Visalu but not against Mukirivuli's descendants is perhaps significant, for it may reflect the importance that the Bashu attach to the incorporation of existing sources of ritual power into the structures of chiefship and thus the impact of Bashu political values on patterns of competition. While Mukirivuli's successors were stronger than Visalu and thus perhaps better able to defend themselves, they also possessed ritual influence over the land that was essential to its well-being and was independent of Luvango's own authority. Thus, if Luvango was to control the land, he had to find a way of incorporating their influence into his chiefship. Visalu's authority, on the other hand, came from the same source as Luvango's, Mukunyu. Luvango, therefore, had little to gain from adding Visalu's authority to his own.

This is not to say that the Babito never employed force against opponents who possessed independent sources of ritual power, for there are cases in which this did occur. For example, Luvango's son Kahese is said to have assassinated Mukumbwa's successor, Mandwa, in Maseki. While Mandwa was a Mubito and, according to his descendants, distantly related to Luvango, his family had established an independent base of ritual power in Maseki and thus played a critical role in the maintenance of the land there. By killing Mandwa, Kahese would appear to have threatened the well-being of this region. It must be noted, however, that Kahese killed Mandwa with the assistance of Mandwa's younger brother and rival, Muvughuli, with whom Kahese allied himself. Following Mandwa's death, Muvughuli took over his brother's ritual duties and performed them in conjunction with Kahese, thus preserving the ritual power of Mukumbwa's family and incorporating it into Kahese's chiefship.[36]

Having defeated Visalu, Luvango proceeded with his own investiture. However, his usurpation had alienated some of his father's supporters who saw Visalu as Mukunyu's legitimate successor. These supporters refused to participate in Luvango's investiture. The most important of the groups that withheld their support from Luvango was the Batangi of Lulinda, who were guardians of Mukunyu's jawbone and regalia and had evidently directed the performance of royal rituals for Luvango's ancestors since Kitara.[37] Faced with the problem of replacing this important family of ritual specialists, Luvango sought the assistance of another family of Batangi who lived in Musindi, to the southwest of the Bashu. The Batangi of Musindi directed royal rituals for the Bamate and several Baswaga chiefdoms. They subsequently sent Luvango a royal drum (*tingi*) and advisors (*baghani*) to teach Luvango's people the procedures to be followed in his investiture.[38]

The participation of the Batangi of Musindi in Luvango's investiture evidently resulted in the introduction of a number of innovations in Bashu royal ritual. This is indicated by a comparison of the terminology and rituals described by individuals involved in the investitures of Luvango's descendants with those involved in the investitures of other Bashu chiefs and with the investiture ceremonies of the Bamate and Baswaga.[39] This comparison suggests that the Batangi of Musindi introduced the use of the terms *baghula* (sg. *mughula*) for the individuals who remove and guard the jawbone of a deceased *mwami*, *semwami* (literally, father of the *mwami*) for the *mwami's* eldest brother, and *semombo* for the person who provides the *mwami* with his *mombo*, as well as certain rituals, including the mock burial of the *mwami* during his investiture, and emblems of authority, such as the above-mentioned *tingi*. Thus political competition within Isale stimulated the process of cultural borrowing. Of perhaps greater importance, however, is the fact that by allying himself with the Batangi of Musindi and adopting these innovations, Luvango moved further away from the pastoral world of his ancestors and closer to the agricultural world of his allies.

By building an extensive network of alliances, undermining the authority of Mukirivuli's descendants, and successfully usurping Visalu's political position, Luvango succeeded in altering the balance of power in Isale
and in establishing himself as a major mountain chief. Luvango's political
strength is perhaps best illustrated by his eldest son's marriage to a woman
whose family (the Bito of Bunyuka) were 'clearers of the forest' in
Bunyuka and important allies of the family of the rainchief Muhiyi.
Luvango's grandfather, Kavango, had previously recognized Muhiyi's
political authority in order to limit Mukirivuli's domination, and
Kavango and his successors had subsequently performed rituals for the
land with Muhiyi's family. It is likely that Luvango arranged, or at least
confirmed, this alliance, for it involved the first marriage of his firstborn
son, and, following Bashu practice, Luvango would have provided the
bridewealth for it. The marriage can thus be seen as an attempt by
Luvango to extend his influence into the area controlled by Muhiyi's
descendants. Whether or not this was his intent, Muhiyi's descendants
evidently perceived it as a threat to their position and reacted to it by
refusing to perform the annual planting ceremonies with Luvango's family, which in effect terminated the alliance that had existed between these
two families since Kavango's arrival in Isale.[40] Since it is unlikely that
Luvango would have risked losing the protection provided by this alliance with Muhiyi's descendants had he not been in a strong position
vis-à-vis Mukirivuli's descendants and other Bashu chiefs, his alliance with
the Bito of Bunyuka reflects Luvango's relative political strength in Isale
at the time of this marriage.

Luvango cemented his alliances by incorporating his allies into the
ritual structures of chiefship, providing each group with specific responsibilities and roles to perform during the accession and mortuary ceremonies. These roles have been described above and are listed in Table 4. In
addition, each group is said to have participated in periodic planting
ceremonies, performing its various roles in conjunction with Luvango's
own sacrifices. Finally, certain of these groups—the Baswaga of Katikale,
the Bashu of Mulamasi, and the Bito of Biabwe—claim that Luvango's
investiture was preceded by the initiation of members of their own family
into their respective ritual offices. These statements, if correct, indicate
that Luvango had begun to expand the temporal as well as the geographical dimension of chiefship, by bringing the life cycles of these leadership
positions into line with that of his own chiefship.

TABLE 4
THE ALLIES OF LUVANGO

Allies	Previous Status	Office
Baswaga of Katikale	*bakonde*, clearers of the forest	royal councillors (*baghani*)
Bito of Biabwe	healers of the land	provided ritual seeds for annual planting ceremonies
Baswaga of Ngulo	rainchiefs, controlled oracle at Musalala	provided *mwami* with iron and copper rings of authority at investiture
Bashu of Mbulamasi	diviners of the land, rainmakers	royal buriers, *baghula* (keeper of royal regalia and jawbone), *semombo*
Bahera of Ngukwe	Senior Bahera settlement	royal councillors
Bashu of Mwenye	dominant clan in Mwenye	—
Bashu of Manighi	clearers of the forest	royal buriers (*bakaka*), provide woman to perform sexual act to cleanse the land following conclusion of mourning for deceased *mwami*
Banisanza of Lisasa	Babito chiefs	—
Mbopi (Matale)	rainchief	*musingya*, directed investiture and provided *mbita*
Bito of Bunyuka (Sine)	clearers of the forest	—
Bahera of Maseki	—	—
Batangi of Masindi	chiefs, royal ritualists among Bamate and Baswaga chiefdoms	*baghani*
Kamesi	Mubito chief in Malio	*mukulu* (*semwami*)

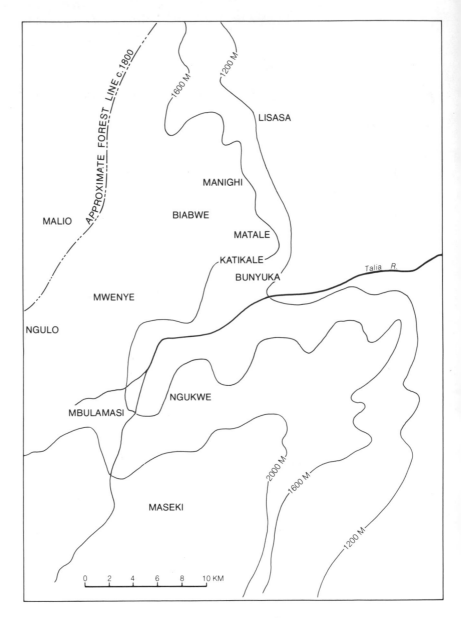

MAP 3. Luvango's Alliances

RITUAL CHIEFSHIP AND THE
DEFINITION OF POLITICAL BOUNDARIES

Having identified Luvango's major allies and rivals, it is useful to step back and view the overall pattern of his political relations. Two important points emerge from this wider appraisal. First, Luvango was clearly selective in his choice of allies. As extensive as his resources may have been, it was impossible for him to form alliances with every group in the Bashu region. Choices had to be made. While the record of his alliances is no doubt incomplete, it suggests that his choice of allies was affected by the need to win the support of local ritual leaders. Thus, while he forged important political alliances with the Bashu of Mwenye and the Bahera of Ngukwe, which were politically influential because of their size, he formed most of his alliances with ritually influential groups. This pattern suggests that Luvango's alliances were directed toward the creation of a wide sphere of ritual coordination, which would permit him to regulate the well-being of the land and to compete with existing rainchiefs. His choice of allies, therefore, conforms to Bashu ideas concerning chiefship and the importance of ritual coordination.

It is, of course, possible that further research may reveal more alliances with groups that possessed little or no ritual influence over the land, and that this would alter our picture of the overall character of Luvango's alliance network. Yet even if this were to happen, it would not alter the fact that ritual leaders formed the core of Luvango's support base, as indicated by their possession of most of the central royal offices associated with Luvango's chiefship (Table 4).

The second point that emerges from this overview concerns the geographical distribution of Luvango's allies. Nearly all of Luvango's alliances were with groups living within Isale or with groups whose support strengthened his position within Isale. In the latter category were the Baswaga of Ngulo, who controlled the water oracle at Musalala, which the people of Isale consulted in making major decisions regarding the land and society; the Batangi of Musindi, from whom Luvango acquired ritual knowledge and regalia necessary for his accession to power in Isale; and the Banisanza of Lisasa, with whom Luvango allied himself in order to reduce the threat of Visalu's challenge to his authority in Isale. The only alliance that does not fit clearly into either category is that with the Bahera of Maseki. As we have seen, this association was an outgrowth of his earlier alliance with the powerful Bahera of Ngukwe and may perhaps be seen as reinforcing his ties to this latter group. The Bahera of

Ngukwe, in fact, claim that while Kahese's mother was born in Maseki, she had been raised in Ngukwe. Within Isale, Luvango's alliances covered both the northern and southern halves of the region.

The question arises as to why Luvango chose to concentrate his alliances within Isale. Why did he not establish more alliances in Maseki or Malio, or extend his influence to the north or further west? One answer may be that he was prevented from expanding his influence beyond Isale by the presence of other powerful chiefs in the areas surrounding Isale. Thus, Luvango may have initially chosen to expand his influence to the west and south because his brother Kamesi was already established in Malio to the northwest, and because the Banisanza of Lisasa, who were expanding their influence into the mountains to the north of Isale, blocked his expansion in this direction. Further expansion to the west may then have been preempted by the presence of the Baswaga chiefs of Ngulo, while further expansion to the south was curtailed by the presence of Mukumbwa's family in Maseki. The political boundaries that separated Isale from surrounding regions and defined the distribution of Luvango's alliances may therefore have been the product of political expediency.

There may, however, have been other factors involved in the definition of these boundaries. To begin with, at the heart of Isale is the Talia River complex, which, as noted in chapter two, created a major break in the rift escarpment and provided an avenue of communications between the mountains and the plains. For Luvango, this communications link may have been critically important, since his political position in the mountains depended initially on his superior access to livestock that were kept in the plains and in the pastures located in the lower valley of the Talia. Expansion away from Isale would have meant moving away from this link to the economic base of his political power. The distribution of Luvango's alliances may therefore reflect his concern for maintaining links to the plains.

A third, more subtle, and yet equally important, consideration may have been the ecological factors that distinguished Isale from the higher and cooler regions of Maseki to the south and the forested regions of Malio and parts of Ngulo to the west and northwest. Isale appears to have differed from both regions as well as from the plains in terms of the distribution of crops and cycles of agricultural activity. It was, in fact, an ecologically distinct zone.

Today variations in crop production between Maseki and Isale are readily apparent as a result of the introduction of wheat, English potatoes, and other European vegetables in Maseki during the colonial era. However, variations in the distribution and cultivation of crops also occurred in the past. While beans and eleusine were produced in both regions, they were planted at different times. In Isale, eleusine was planted with sorghum before the long autumn rains and was harvested in December. Beans were planted along with peas and some maize before the short spring rains in February and harvested in June and July. In Maseki and the higher elevations to the south, this rotation was reversed. Eleusine was planted along with colocassia before the spring rains, while beans were planted in the fall. These variations reflect differences in regional rainfall patterns. The lighter rains needed for growing beans fall in the higher areas to the south between September and November, whereas the heavier rains required for growing eleusine fall in the spring.[41]

Isale was evidently also distinguished from eastern Malio and parts of northern Ngulo, which, as noted above, were still heavily forested at the time of Luvango's expansion in Isale and probably supported an economy —similar to that which is still practiced in the forested regions of western Malio—in which hunting and gathering were combined with agricultural activities centering primarily on the production of bananas.

Ecological distinctions between Isale and its neighbors may have contributed to the definition of Isale's political boundaries and to the distribution of Luvango's alliances not so much because they created physical barriers to political expansion, though this may have been partially true in the case of Malio, but because they set limits to the coordination of ritual influence over the land and thus to the expansion of ritual chiefship. Within Isale, because of its ecological homogeneity, planting rituals, invocations for rain and dry weather, and the cleansing of the land could be coordinated to occur at the same time. Moreover, because major natural disasters often affected the region as a whole, this coordination was evidently perceived as important for the maintenance of the land. Ritual coordination between Isale and neighboring regions, on the other hand, while desirable in the face of particularly widespread disasters, was difficult to achieve because of differences in the distribution of crops, the timing of agricultural cycles, and the nature of ritual observances. For example, the annual planting ceremonies marking the beginning of the agricultural year preceded the planting of eleusine in both Maseki and Isale, because eleusine was said to be the first crop, in the sense of being the oldest. However, eleusine was planted in the fall in Isale and in the spring in Maseki because of differences in rainfall. This made the coordination of new year planting ceremonies difficult and may have inhibited

the ritual and thus political unification of the two regions. The importance of this obstacle was highlighted at the end of the nineteenth century, when Luvango's grandson Vyogho succeeded in expanding his
political authority into Maseki. As we shall see in chapter seven, this
expansion was accompanied by changes in the timing of the new year
ceremonies in Maseki, designed to coordinate them with those in Isale.

Ritual coordination between Isale and the forested regions to the
northwest may also have been difficult to achieve because of cultural
differences between these areas. The forest environment and the dominance of forest Bantu peoples in Malio and northern Ngulo created
differences in the nature of ritual practices related to the land, differences
that are still observable today because of the relatively recent occupation
of Malio and northern Ngulo by Nande cultivators. Thus, the central
fertility rites in Malio center on invocations to the serpent spirit
Mulumbi, which is presently guarded by the people of Mulali. While
Mulumbi is also found in Isale, it is associated with groups who are
recognized as the descendants of former forest dwellers and is of secondary importance to the ritual protection of the land. Conversely, the
distribution of ritual seeds (*mikene*), which is associated with seed agriculture and is the focal point of the annual planting ceremonies in Isale, is
of only secondary importance in Malio today and appears to date from
the clearing of the forest. Because of these differences, ritual coordination
between Isale and the forested areas to the northwest may have been
impeded. This in turn may have inhibited the political integration of these
areas within the context of Bashu chiefship.

The distribution of Luvango's alliances and the limits of ritual authority may thus have been defined by a combination of factors, including
the presence of existing chiefs, the need to maintain links to the plains,
and difficulties in coordinating ritual activities beyond the borders of
Isale. Within these limits, however, Luvango created a coordinated sphere
of ritual control, which unified much of Isale and gave it a political
identity that has survived until the present day. Thus, Luvango's successors continued to view Isale as a unified territory over which ritual
influence should be coordinated.

CONCLUSION

From the perspective of the people of Isale the establishment of Luvango's chiefship marks a watershed in their political history. Before this
event, the people of the region generally claim that there were no true

chiefs in Isale. Thus, despite the cultural and functional similarities between earlier rainchiefs and the present Babito chiefs, the two forms of leadership are viewed as categorically distinct. This distinction is, I suggest, based on Bashu perceptions concerning the relative ability of each type of chief to control the well-being of the land. The Bashu claim that, while rainchiefs could control the rains and help insure a good harvest, they were unable to control totally the well-being of the land. Luvango and his descendants, on the other hand, possessed the power to do so.

This evaluation is not simply a product of Bashu deference to their present chiefs. Nor does it reflect the absence of serious famines during the period of Babito rule. Major climatic crises occurred following Luvango's death and at the end of the nineteenth century. Moreover, there is evidence that famine conditions occurred during Luvango's reign.[42] The distinction is based instead on Bashu perceptions of the political transformation achieved by Luvango and maintained by his successors.

From a Bashu viewpoint, rainchiefs were less effective in controlling the land because their spheres of ritual influence were limited, seldom extending over more than two or three adjacent ridges, whereas major climatic crises frequently affected a wider region—often Isale as a whole. To end such crises rainchiefs had to consolidate their ritual authority. Thus, Mukirivuli's descendants claim that their family occasionally cooperated with that of Muhiyi during times of regional famine. Yet wider coordination was difficult to maintain during periods of plenty, when each family attempted to increase its sphere of influence and competed for the support of local ritual leaders. Thus, famines were difficult to prevent.

Luvango, through his superior access to economic resources, was able to construct a much wider sphere of ritual influence bringing most of Isale under the control of a single chief. I suggest that from a Bashu viewpoint, this enlargement of political scale increased the possibility for ritual control of the land, permitting Luvango to 'cover the land' and thereby to domesticate the forces of nature, where his predecessors had been unable to do so except in times of crises. While this did not eliminate the possibility of famine, from a Bashu point of view it decreased its likelihood. Moreover, when famine did occur, the sphere of influence established by Luvango enhanced his credibility and evidently enabled him to provide an explanation for the crisis that allowed him to sustain his ritual legitimacy. For Luvango was able to maintain his chiefship despite the apparent occurrence of famine conditions during his reign.

The importance of wider spheres of ritual influence for maintaining political legitimacy during periods of famine will be examined in detail in the next chapter.

I suggest, therefore, that it was not simply the enlargement of political scale achieved by Luvango that accounts for his success in establishing control in Isale; it was also Bashu perceptions of what this political transformation meant in terms of the world in which they lived. Luvango's political success in Isale resulted from his ability to adapt his political actions to conform with Bashu ideas about the nature and function of chiefship as much as from his access to economic resources and his success in building alliances. In this way, Bashu political values helped shape the processes of domination and state formation in Isale.

CHAPTER FIVE

SUCCESSION, FAMINE, AND POLITICAL COMPETITION IN ISALE

The period following Luvango's death, from approximately the 1860s to the 1880s, saw a fragmentation and proliferation of Babito chiefship in Isale. The period began with a succession struggle between two of Luvango's sons, Muhoyo and Kivoto. While this competition was initially won by Kivoto, Muhoyo succeeded in establishing his own chiefship in northern Isale, thereby dividing Babito rule. Meanwhile, Luvango's firstborn son, Mutsora, established his authority over the central Vusekuli area of Isale, while the descendants of Luvango's brother Visalu took advantage of the divisions among Luvango's sons to reestablish their claim to chiefship in the Mughulungu area. As a result of these developments, four separate and largely autonomous Babito chiefdoms emerged in Isale.

Succession disputes leading to the political fragmentation of a ruling lineage and the creation of a cluster of related and yet independent states, what A. Richards (1961: 144) has termed a "multi-kingdom tribe," constituted a common pattern of political development in precolonial eastern and southern Africa.[1] This pattern was particularly common in societies in which bureaucratic relations were kinship-based, and where succession rules were unclear, or, as in the Bashu case, defined in such a way as to open the succession to a number of potential candidates. Examples of the phenomenon can be seen in the histories of chiefly expansion among the Zande, Soga, Nyanga, Hunde, Alur, Haya, Sukumu, Bemba, Xhosa, and Pedi.

While this pattern of political development has a wide distribution, the specific form that it took in any given society, as well as the nature of subsequent relations among member units of the resulting cluster of related states, was, I suggest, colored by local conceptions of the nature and function of political leadership. Thus, as will be seen in the present chapter, the events that led to the fragmentation and replication of Babito chiefship in Isale, and the relations among the resulting chiefdoms, revolved around the problem of ecological control, which as we have seen, was at the heart of Bashu political values. Like the process of domination, subsequent competition among the Babito involved attempts by candidates for chiefship to create spheres of ritual influence in Isale while undermining those of their rivals. The outcome of these attempts and the relative success of each competitor were ultimately defined by the onset of a major famine, which, as suggested earlier, tested each competitor's legitimacy. In these ways, the widespread phenomena of royal lineage segmentation and political fragmentation were, in the Bashu case, shaped by the ideology of ritual chiefship.

SUCCESSION AND FAMINE IN NORTHERN ISALE

Bashu succession practices were closely tied to ideas about the nature and function of chiefship. In principle, the *mwami* was always succeeded by the firstborn son of the *mombo*, a special wife he received at his investiture. In actual practice, however, the *mombo's* maternity was social rather than biological, and was ascribed to whichever of the deceased *mwami's* sons was chosen by a special council of investors, the *basingya*. While this choice is said to have reflected the deceased *mwami's* wishes, it appears that the real decision-making power lay solely with the *basingya* itself, since the *mwami's* wishes were only made known through the *basingya's* announcement of who would succeed as chief. The succession was thus open to any of the preceding *mwami's* sons and was determined by competition among these sons for the support of the *basingya* rather than by fixed rules of succession. Only the firstborn son of the *mwami's* first wife, who was designated *mukulu*, was in principle barred from this competition and the succession.

From a Bashu perspective this system insured not only popular support for the new *mwami* but also the renewal of the land. For the *basingya* was made up primarily of important ritual leaders and former chiefs who were collectively responsible for the performance of land rituals. By gaining the support of the *basingya*, a candidate for chiefship, in effect, reconstructed the sphere of ritual influence established by his predecessor, and thereby 'covered' the land and insured its continued productivity.

Competition for the support of the *basingya* could in theory begin only after the death of the preceding *mwami*. For it was thought that competition during the preceding *mwami's* reign would endanger the well-being of the land by disrupting the network of alliances needed to control the land and prevent disaster. Thus, while candidates for chiefship may have engaged in some maneuvering before their father's death, open competition only began subsequent to this event. This meant that the interregnum period was often extended, lasting in some cases several years. While this practice was potentially disastrous, in that it opened up the land to misfortune, it was necessary in light of the need for ritual unity and the prohibition against competition during the preceding *mwami's* reign.[2]

Given the Bashu system of succession, it is not surprising that Luvango's death was followed by a succession struggle between two of his sons, Kivoto and Muhoyo. This competition was initially won by Kivoto, who was invested as Luvango's successor. However, Kivoto appears to have lacked extensive support among Luvango's former allies in northern Isale, where, following tradition, he established his royal compound near his father's home at Mutendero. This lack of support eventually resulted in his overthrow and in the investiture of Muhoyo as *mwami* in northern Isale.

Kivoto's support base lay primarily to the south, where, along with his mother's brothers, the Bahera of Ngukwe, his major allies were the Bashu of Mbulamasi and his eldest brother, Mutsora, who had been invested as Kivoto's *mukulu* and therefore had a vested interest in his chiefship. In northern Isale, Kivoto had few supporters aside from the Bito of Biabwe, who were Mutsora's maternal relatives.[3]

The Baswaga of Katikale, whose status as 'clearers of the forest' at Mutendero made their cooperation in land rituals essential, joined with Muhoyo's maternal relatives, the Baswaga of Ngulo, who refused to participate in Kivoto's investiture, and opposed Kivoto's authority.[4] The people of Mwenye, whom Luvango had assisted in their fight against the Basumba (see chapter four, p. 97), also supported Muhoyo in opposing Kivoto. Muhoyo's marriage to the daughter of Kanyabwarara, whom Luvango had placed in Mwenye as his representative, may have affected this decision.[5]

The rainchief Mbopi, who had presented Luvango with his *mbita*, also refused to participate in Kivoto's investiture and eventually provided Muhoyo with his *mbita*. Mbopi had no overriding kinship tie to Muhoyo that would account for this support. However, it will be remembered that Mbopi's father dominated northern Isale when the Babito arrived and that his descendants were subsequently eclipsed by Luvango. Mbopi's decision to support Muhoyo may therefore have been based on a desire to strengthen his family's position in Isale by contributing to a division in Babito authority. In response to this decision, Kivoto acquired the services of Kakuse, a rainchief from Maseki, to fill the ritual role abandoned by Mbopi. This arrangement solved the problem of having an invested chief provide Kivoto with his *mbita*. It did nothing, however, to strengthen Kivoto's position in northern Isale.[6]

The Bashu of Manighi, who were Luvango's royal buriers *(bakaka)* as well as important 'clearers of the forest' in northern Isale, also refused to support Kivoto. This decision may have been affected by their long-standing relationship with Mbopi's family, to whom they were also *bakaka*. Of perhaps greater importance, however, was the fact that they improved their status in Isale by supporting Muhoyo. While they had been royal buriers for Luvango, they acquired the more prestigious position of *baghula* under Muhoyo, replacing the Bashu of Mbulamasi, who chose to support Kivoto. Muhoyo may therefore have won the support of this group by promising them this position and thereby increasing their vested interest in his chiefship.[7]

Kivoto's position among his father's former allies in northern Isale was, therefore, weak in comparison with that of Muhoyo. To make matters worse, Muhoyo strengthened his position in northern Isale by reestablishing the alliance that his great-grandfather, Kavango, had created with the family of the rainchief Muhiyi. This alliance had been broken by Muhiyi's descendants in response to Luvango's attempt to subvert their authority (see chapter four, p. 100).[8]

Problems in determining the precise timing of particular alliances make it difficult to know whether Kivoto's loss of support in northern Isale occurred before or after his investiture. But one would assume, given the importance the Bashu attach to maintaining the cooperation of ritual leaders, that the *basingya* would not have invested Kivoto had he not had substantial support among its membership. On the other hand, two of Muhoyo's supporters, the Baswaga of Ngulo, and Mbopi, claim to have refused to participate in Kivoto's investiture. While these claims may represent retroactive attempts to increase their status as long-term allies of Muhoyo's family by pushing back the time of their initial support, there is evidence that in at least one case, that of Mbopi, Kivoto was forced to acquire the services of a new family of chiefs to provide him with his *mbita*. It is thus possible that the *basingya* were split over whom to invest. The southern *basingya*, led by the Bashu of Mbulamasi, who possessed Luvango's jawbone and other regalia needed for the investiture, may then have gone ahead and invested Kivoto on their own, leaving Muhoyo with substantial support in northern Isale but without the em-

TABLE 5
THE ALLIES OF KIVOTO AND MUHOYO

Kivoto's Allies	Status	Office
Bito of Biabwe	healers of the land	owner of ritual seeds (*mikene*)
Bashu of Mbulamasi	diviners of the land and rainmakers	guardians of royal regalia (*baghula*), *semombo*
Bahera of Ngukwe	senior Bahera settlement	*semombo*
Mutsora (Vusekuli)	eldest son of Luvango	*mukulu*
Kakuse (Maseki)	rainchief	*musingya*, presented Kivoto with *mbita*
Batangi of Lulinda (support acquired after famine)	royal ritualists for early Babito chiefs	directed ritual activities for Kivoto and his son Vyogho
Basyangwa of Muluka	clearers of the forest	?
Muhoyo's Allies		
Baswaga of Ngulo	rainchiefs, controlled oracle at Musalala	*musingya*, provided iron and copper rings (*miringa*); *semombo*
Baswaga of Katikale	clearers of the forest	royal buriers (*bakaka*)
Bashu of Manighi	clearers of the forest	guardians of royal regalia (*baghula*)
Descendants of Muhiyi (Bunyuka)	rainchiefs	royal councillors (*baghani*)
Mbopi s/o Mukirivuli (Matale)	rainchief	*musingya*, presented Muhoyo with *mbita*
Bashu of Mwenye	dominant lineage in Mwenye	?

blems and sources of authority needed to establish his claim to Luvango's chiefship. The difficulty with this explanation is that it does not account for the fact that Kivoto was able to establish his homestead at Mutendero in northern Isale.

Perhaps the most likely possibility is that Kivoto possessed some support in northern Isale at the time of his investiture, which gave him an overall majority among the *basingya*. However, subsequent to his investiture, Muhoyo succeeded in parlaying his own support among his moth-

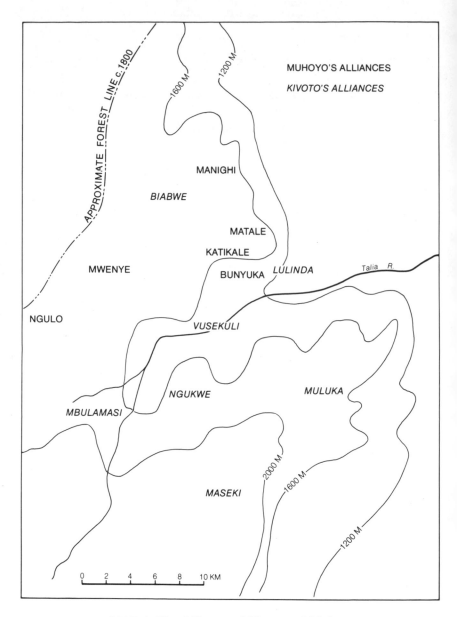

MAP 4. The Alliances of Kivoto and Muhoyo

er's brothers, the Baswaga of Ngulo, and with Mbopi, both of whom claim to have refused to participate in Kivoto's investiture, into wider support among their allies, the Baswaga of Katikale and Mbopi's *bakaka*, the Bashu of Manighi. This interpretation would explain why Kivoto was initially able to establish his royal homestead in northern Isale, even though certain groups in the region refused to participate in his investiture.

Whichever explanation one accepts, it is clear that Kivoto failed to establish widespread support in northern Isale. This lack of support meant that while he was apparently able to maintain his claim to chiefship during years of successful harvests, he was unable to do so when several years of abnormally heavy rains brought on a famine.

Before examining the events surrounding this famine, we must deal with the historiographical problem of determining whether or not a famine actually occurred at this time. Climatic crises are frequently associated with periods of political disruption in Bashu historical traditions, as they are in the traditions of many African societies. This correlation may reflect the impact of natural events on societies in which political leadership is tied to the problem of ecological control. It is possible, however, that given this ideological association, some references to famine are culturally determined characterizations of periods of political disruption and not references to actual natural disasters. One must, therefore, be careful in accepting the historicity of such references.

While there are no eyewitness accounts to verify the existence of the famine that is said to have occurred following Luvango's death, there is indirect evidence that suggests that it did occur. First, the traditions that describe the famine are not simply stereotypic references to famine but detailed accounts of what happened during the famine. This detail suggests that the famine is more than a cultural metaphor.[9] Second, while conditions of famine and plenty do serve at times as cultural metaphors for opposing political conditions in Bashu traditions, ideas about cycles of chiefship and the changing fertility of the land have not completely dominated Bashu historical memory. Thus, as noted in chapter four, there are references to famine having occurred during the reign of Luvango, despite his great success in consolidating ritual control over the land.

Third, this particular famine occurs in the traditions of three regions of Isale and is associated with three separate sets of events. In the Mutendero area it is linked to the competition that occurred between Muhoyo and Kivoto. In Vusekuli it explains how Mutsora overcame the region's former chiefs, while in the Mughulungu area it is associated with the

establishment of Visalu's descendants as chiefs in Isale. While these traditions describe different sets of events, there is evidence that they refer to
a single famine. First, the events described in the traditions are contempo-
raneous. Second a variant of the Vusekuli tradition connects the famine
in Vusekuli with the succession struggle described in the Mutendero
traditions, indicating that both traditions refer to the same famine.

> There was a *mwami* who had three sons [a reference to Luvango and
> his three sons, Mutsora, Kivoto, and Muhoyo]. One of them received
> the power [*bwami*]. However, one of the other two [Muhoyo], who
> was discontented and evil, brought on a famine. At this time the people
> went to Mutsora and asked him what they should do. He told them,
> ... and after that there was a good harvest.[10]

Fourth, the traditions of all three areas agree that the famine was caused
by a period of excessive rains. These traditions therefore appear to describe separate local reactions to a single famine and provide independent
testimonies to its existence. This would not be so if the traditions of the
three areas referred to the same events, for then they would be variants
of a common tradition and provide only a single testimony to the famine's
existence.

It should also be noted that in the Vusekuli region the famine and
Mutsora's role in stopping it provide the justification for the expansion
of Babito authority into the area. In general, the Babito base their right
to rule on first occupancy. Here, the first occupancy claim is forsaken
in favor of the famine tradition, which admits to the existence of prior
chiefs in the region. This is not only exceptional, it is, to the best of my
knowledge, the sole exception, and I suggest that if the famine had not
actually occurred, and if Mutsora's political success was not tied to it, this
exception would not exist. In addition, the name Mutsora means 'savior'
in Kinande and is said to have been given to Mutsora following the
famine.[11]

Finally, the attribution of the famine to a long period of excessive rains
is consistent with climatic data for East Africa, which indicates that this
period, the 1860s and 1870s, was much wetter than normal (Nicholson,
1976: 147). There is thus good reason to accept that the famine associated with this period in Bashu traditions actually occurred.

The events surrounding the famine in Mutendero are described in the
following tradition:

> Long ago the people here planted their eleusine and beans. Then some
> people came from Buswaga. They had a string of *vigholero* and also of

syongwangi [two types of beads]. They also had a goat. After that there was much rain and all the crops rotted. Even the bananas. People began to eat the trunks of the banana trees. This happened after the death of Luvango, and the people decided that they should call Muhoyo, who had gone to Ngulo, from where his mother had come. They sent some people with five goats to bring him back to Mutendero, where he performed the sacrifice and ended the famine. Because of this the people said that Muhoyo sould be the *mwami*. [12]

Other traditions state that Kivoto tried to make his own sacrifice to end the rains but failed, and the people of the region then turned to Muhoyo, forcing Kivoto to flee from Mutendero and seek refuge at Tutu and later at Vuhovi, near Mutsora's settlement in Vusekuli. It thus appears that, having failed to establish widespread support in northern Isale, Kivoto was unable to maintain his legitimacy as a chief when faced with an extended climatic disaster. His ritual credibility was quickly exhausted, and, rather than acknowledge Muhoyo's superior authority, he chose to leave the area and settle further south, where his support was stronger.

Kivoto settled in Vuhovi, where he was welcomed and given land by his eldest brother, Mutsora, who, as described below, had succeeded in establishing his authority over Vuhovi during the famine. Mutsora recognized Kivoto's right to Luvango's chiefship and lent him his support along with that of the people of the Vusekuli area. [13]

Having settled at Vuhovi, Kivoto strengthened his position in southern Isale by forming two important marriage alliances. The first was with the Basyangwa of Muluka, who were clearers of the forest in Muluka, neighboring Vusali, and parts of Maseki. A branch of this family also claimed the power to make rain. The second alliance was with the Batangi of Lulinda, who had directed the performance of royal rituals for Kivoto's ancestors since Busongora but had refused to recognize Luvango's descendants as legitimate chiefs, choosing instead to support Visalu and his heirs. While the events leading up to this shift in Batangi support will be discussed in chapter seven, it should be noted here that Kivoto, or, perhaps more accurately, his *basingya*, solidified this important alliance by choosing Vyogho, the son of Kivoto's Mutangi wife, as his successor. Thus Kivoto was able to maintain and strengthen his political position in southern Isale, thereby dividing Isale in half. [14]

The north-south division in the distribution of Muhoyo's and Kivoto's alliances may have been fortuitous. However, it may also have reflected a more fundamental social division within Isale. As will be seen in chapter eight, a similar division of Isale occurred during the colonial era following the reunification of Isale under Kivoto's son Vyogho at the end of the nineteenth century. This split occurred between two branches of Kivoto's descendants and thus did not simply represent the reemergence of tensions between the descendants of Kivoto and Muhoyo. The north-south division of Isale may therefore have been based on other factors.

While there is no indication that major economic or cultural differences existed between the two halves of the region, the valley of the Talia River forms a deep divide between northern and southern Isale. This divide may have inhibited social interaction between the two halves of the region and contributed to the development of separate networks of social relationships within each area. Thus the somewhat fragmentary evidence of marriage alliances between commoner groups in Isale indicates that most, though not all, marriages fell within one or the other of, rather than between , the two halves of the region. Similarly, while the people of northern Isale visited the Bito of Biabwe to receive divinations for the land, the people of southern Isale consulted the Bashu of Mbulamasi.

This social division may have helped shape the distribution of Kivoto's and Muhoyo's alliances. Each prince was raised among his maternal uncles, who subsequently became his most important allies. It was to these maternal uncles that a prince looked for advice and protection throughout his life. It was only natural, therefore, that a prince's choice of allies would be influenced by the social connections that his maternal uncles possessed. The degree to which these connections were limited to either northern or southern Isale would then serve to limit the subsequent distribution of the prince's own alliances to one or the other region. Thus, the social relations of Kivoto's maternal relatives, the Bahera of Ngukwe, who lived in southern Isale, may have facilitated and encouraged the concentration of his support-building activities in this half of the region. In this connection, it is perhaps significant that the Bahera of Ngukwe were related to Kivoto's other major ally in southern Isale, the Bashu of Mbulamasi, through several marriages. In addition, while I could find no evidence of marriages between either of these groups and Kivoto's later allies, the Basyangwa of Muluka and the Batangi of Lulinda, the Basyangwa were connected to the Bashu of Mbulamasi through their reliance on the people of Mbulamasi for divinations related to the land. Muhoyo's allies can be linked to the social networks of his maternal uncles, the Baswaga of Ngulo, in a similar fashion. Thus, the Baswaga of Ngulo were related by common descent to the Baswaga of Katikale, who were themselves allies of the family of the rainchief Mukirivuli, whose royal ritualists were the Bashu of Manighi. The dependence of each prince on his maternal uncles' connections, combined with the tendency for these connections to be centered in one-half of Isale may therefore have contributed to the resulting north-south division of Isale following Luvango's death.

The famine that followed Luvango's death can thus be seen to have defined the actual distribution of ritual influence in Isale and, in doing so, to have divided the region between Muhoyo and Kivoto. At the same time, the famine provided their eldest brother, Mutsora, with an opportunity to carve out his own sphere of influence in the central Vusekuli region of Isale, while to the east, their cousin Molero, son of Visalu, made use of the famine to establish his own authority over the Mount Mughu-lungu area of Isale.

FAMINE AND POLITICAL EXPANSION IN ISALE

The Vusekuli area of Isale, located in the valley of the Talia River between Vuhovi in the north and Mbulamasi to the south, is one of the oldest areas of Nande settlement in the Mitumba Mountains and, more importantly, the center of the oldest chiefship in Isale. The chiefs of Vusekuli trace their ancestors back to the period of early Nande settlement, though not necessarily in an unbroken line of descent, since there is evidence that the chiefs of Vusekuli at the time of Babito expansion into the area had themselves replaced an earlier line of chiefs.

The political and ritual authority of the Vusekuli chiefs at one time extended into nearly all the areas immediately adjacent to Vusekuli. Thus, groups from Mbulamasi, Mutendero, Vuhovi, Kahondo, and Muvulia have traditions of having formerly taken tribute in the form of *ngemu* to the chiefs of Vusekuli, in recognition of the Vusekuli chiefs' important role in the performance of land rituals and in the making of rain in the regions surrounding Vusekuli. These payments are, in fact, still made, though today the chiefs of Vusekuli pass them on to the Babito.[15]

Because of their geographical position in the center of Isale and the distribution of their ritual influence to the north and the south of the Talia River, the chiefs of Vusekuli represented a potential political bridge between northern and southern Isale, and thus an important link in the establishment of a unified sphere of ritual influence over Isale. It is therefore not surprising that the Babito would wish to extend their political authority over Vusekuli and incorporate the ritual influence of its chiefs into their own spheres of influence, especially given the apparent social division that separated the two halves of Isale. It was perhaps with this in mind that Luvango's firstborn son, Mutsora, chose to settle in Vusekuli.

Mutsora's descendants claim that Mutsora settled in Vusekuli following his father's death and that he was invited there by Ghotya, the *mwami* of Vusekuli, who hoped that Mutsora, as Luvango's son, would be able to end the rains that were causing famine.[16] The traditions of other groups living in and around Vusekuli, on the other hand, place Mutsora's arrival *before* the famine and indicate that he undermined Ghotya's authority and subsequently replaced him as chief when Ghotya was unable to stop the rains.[17] The discrepancy between these commoner traditions and that of Mutsora's descendants may indicate that Mutsora's descendants have moved up the time of Mutsora's arrival in Vusekuli in order to eliminate any suggestion that he undermined Ghotya's position. If so, Mutsora's settlement in the area may have preceded Luvango's death and been part of Luvango's wider strategy in Isale. The political importance of Vusekuli, combined with Luvango's apparent attempts to unite the two halves of Isale under his ritual authority, make it somewhat surprising that he would not have attempted to establish his authority in Vusekuli.

If, on the other hand, Mutsora's descendants are correct in placing his arrival in Vusekuli after Luvango's death, he may have been operating on his own, though it would appear more likely that he was an agent for Kivoto. Mutsora had, after all, acknowledged Kivoto's right to succession and participated in his investiture. In addition, while Mutsora and his successors retained considerable autonomy in Vusekuli, they were never invested as chiefs but continued to support the chiefship of Kivoto's lineage. Mutsora may therfore have settled in Vusekuli in order to strengthen Kivoto's position in Isale, as well as to increase overall Babito control over the region.

Whatever the precise motivation behind Mutsora's settlement in the Vusekuli area may have been, he did not initially settle in Vusekuli itself but chose instead to settle among some of Ghotya's subjects *(basoki)*, the Bakira of Kirungwe, whose settlement was located to the east of Vusekuli, in the lower valley of the Talia. Mutsora evidently formed an important alliance with this group, for they are mentioned in nearly every tradition that describes Mutsora's actions during this period. The exact nature of the alliance and of their support, however, is unclear, for the details are no longer remembered by Mutsora's descendants, and I was unable to locate a descendant of this section of the Bakira clan, who had been forced to evacuate their settlement at Kirungwe on account of sleeping sickness during the early colonial period and had for the most part immigrated to Uganda.[18]

According to Ghotya's descendants, the Bakira sent word to Ghotya that a stranger had arrived among them. Ghotya then went to see who the stranger was.

> When he arrived there he saw that it was Mutsora, the son of Luvango, and therefore returned home to get some beer. He then returned to Kirungwe and welcomed Mutsora and gave him a parcel of land at Lulika [near Vusekuli].[19]

A variant of this tradition suggests that Ghotya was not happy having this influential person living among his *basoki* and insisted that he come and settle near Vusekuli, presumably so that he could keep an eye on him.

> Several Bakira went to tell Ghotya that Mutsora had arrived among them. Ghotya began to prepare beer and sent three calabashes to Kirungwe, saying that Mutsora must come and settle at his place.[20]

The traditions agree, however, that Ghotya initially welcomed Mutsora and that they joined in performing sacrifices for the land.

> When Mutsora arrived in Lulika, Ghotya called together his *basoki bakulu*: Kyambu, Kahindolya, Vutundya, Vulema, Ngologo, and Nyaviremba. Ghotya presented Mutsora to them and told them that Mutsora could bring much food and that he was the son of the *mwami*. Ghotya gave Mutsora the territory of Lulika. As a result of this, much food was produced and Ghotya was content.[21]

Another tradition states:

> ... when there was a sacrifice, Ghotya made the first sacrifice and then Mutsora made his.[22]

The traditions of this region also agree in suggesting that Mutsora slowly increased his influence in the region. This is indicated in the first of the two preceding traditions by the following lines: "Ghotya again gathered his *basoki* and told them that from then on they were to take their *mubako* to Mutsora and that Mutsora would take it to Ghotya."[23] The fact that this arrangement was made at a second meeting suggests that Mutsora's influence increased after he settled in Vusekuli. The expansion of Mutsora's influence in also indicated in the traditions related by Ghotya's descendants. In these, Ghotya is said to have given Mutsora land at Lulika, and "Later, because Mutsora had many goats and cattle and was afraid that they might fall in the [Talia] river, Ghotya gave him the place called Katiri."[24] Thus Mutsora is shown to have extended the area over which he had direct control. The reference also suggests how he accomplished this, for the claim that Mutsora kept his herds at Katiri does not fit with the ecology of the area. The valley of the Talia rises above 1500 m and is quite narrow at this point. For both reasons, the area is not well suited for herding plains cattle. On the other hand, if one moves a short distance east along the valley, the river descends toward the plains and the valley widens, providing what were evidently good grazing lands, for it was here that Mutsora first settled and that his nephew Vyogho is also said to have kept herds. Thus, while Mutsora may have kept some goats near his homestead at Kitiri, it is likely that the major portion of his herds were kept to the east. Consequently, the statement that Ghotya gave Mutsora more land because he had too many animals may be a reference to Mutsora's general wealth in livestock and to the fact that it was this wealth in livestock that allowed him to expand his influence, rather than a description of actual events.

Mutsora, like other Babito, presumably used his wealth in livestock to create marriage alliances with other groups in Vusekuli. However, I was unable to verify this fact with many specific examples. This difficulty needs some explanation, given the general wealth of such data among the Bashu. During the early 1960s, the authority of Mutsora's descendants in Vusekuli was seriously challenged by local rivals, a situation that was exacerbated by the changes in the national government and by the arrival of the Simba Rebels in 1964. The intensity of feelings that still surrounds these events not only prevents my detailing them here but also hampered my attempts to solicit specific information concerning Mutsora's alliances in Vusekuli. In order to solidify their claim to chiefship, Mutsora's descendants maintain that all of Mutsora's supporters were members of his own lineage. While this hardly seems likely, given the general pattern of support-building elsewhere in Isale, it prevents Mutsora's descendants from providing marriage data, for evidence of marriage between Mutsora and his clients, given rules of lineage exogamy, would undercut this claim. The reluctance of other groups to supply this information apparently stems from the fact that the challenge to the position of Mutsora's descendants was ended with the arrest and mysterious death of the man who had been the center of this challenge. As a consequence of these difficulties, I was able to collect data only on Mutsora's first marriage to a woman from Bunyuka, which, as noted in chapter four, appears to have been part of Luvango's attempt to expand his influence into this region. It had little direct effect on Mutsora's position in Vusekuli, since there were only a few Bito living in the area, and none of these appears to have had any ritual or political importance.

There is evidence within Vusekuli, however, that Mutsora did succeed in winning the support of Ghotya's most influential subjects, the family of Kahindolya, whose ancestors were the original chiefs in Vusekuli and whose cooperation was thus critical to the maintenance of the land. Kahindolya also possessed direct control over the area of Vutungu, to the west of Vusekuli.[25]

Kahindolya's descendants claim that Ghotya had used force to steal their power over this region before Mutsora arrived, and that they had, therefore, allied themselves with Mutsora in order to weaken Ghotya's power. Their opposition to Ghotya is reflected in their version of the story of Mutsora's arrival.

> After Ghotya had settled there, the people of Kirungwe sent word one day that there was a stranger among them. Ghotya went to see who it was and found that it was Mutsora. Ghotya then returned and got some beer which he took to Kirungwe. At Kirungwe he drank the beer with Mutsora. Ghotya, however, became drunk and he attacked and killed someone. The people then became angry and chased Ghotya away.[26]

Since this is the only version in which this elaboration of the beer-drinking episode occurs, this version must be seen as an effort to discredit Ghotya.[27]

Mutsora's authority was also recognized by the Basumba of Lukanga. However, the descendants of this group suggest that they were pressured into accepting Mutsora's family as chiefs. Thus their submission may have postdated the famine. The Basumba of Lukanga subsequently provided Kivoto and his son Vyogho with their ritual services, relighting the royal fires at Vyogho's investiture.[28]

Whatever other alliances Mutsora might have made, the support of Kahindolya, combined with his prestige of being Luvango's firstborn son, placed him in a strong position to challenge Ghotya's authority during the extensive rains that hit the region at this time. A number of traditions claim that the people of Vusekuli went to Ghotya when the famine began and asked him to drive the famine away with his sacrifices. When Ghotya failed to end the famine, the people turned to Mutsora for assistance.[29]

Some informants claim that Mutsora ended the famine simply by performing the necessary sacrifices. Others provide a more detailed account of Mutsora's actions and, in doing so, present a clearer picture of how Mutsora achieved his success and of how, in general, Bashu chiefs succeeded in convincing their followers that they could control climatic change despite the onset of natural disasters. According to these informants, Mutsora told the people of Vusekuli

> . . . that they should tear down all of the old houses. They should then wait to see if plants grew from the remains of the destroyed houses. If they did, they should not eat these plants but wait until they were ripe and then take the seeds from them and plant them in the fields. These seeds would produce a bountiful harvest.[30]

This advice exhibits a remarkable blend of cultural and practical knowledge, combined to create what, from an external viewpoint, was a political bluff. The advice plays on Bashu customs associated with the tearing down of huts. When an old house was torn down because it was falling apart, a new house could not be constructed on the same location until plants had germinated in the soil of the remaining mound, for their growth was a sign that the ancestors who had been sacrificed to in the old house were satisfied with the move to the new location (the new house was never built in exactly the same spot) and were in general contented.

This practice was based on the general premise that ancestors could affect the productivity of the land. Thus if they were angry they could cause the land to be unproductive and prevent the plants from growing in the house mound. Conversely, if they were contented, the land would be fertile and the plants would grow in the mounds. As a result of these concepts, house mounds served as an explanatory mechanism for famines, for, given the richness of the growing medium provided by a house mound, and the near certainty that some grains of millet, sorghum, or beans remained in the debris from the hut, it was rare that plants did not reproduce in these mounds. In fact, the only time when nothing grew was when there was extreme weather and the land in general was unproductive. At such times the house mounds provided an explanation for the bad weather, i.e., the ancestors were angry. This, however, was only half of an explanation. The people then had to determine why the ancestors were angry, and it was here that competing chiefs entered the process, each explaining the anger of the ancestors in such a way as to deflect ultimate responsibility for the disaster and to maintain their credibility. If the mounds failed to produce plants, a chief could indicate that this was because the ancestors were angry at a rival, in this case Ghotya, for opposing him and disrupting ritual control over the land or for directly offending them. If, on the other hand, plants grew in the house mounds, the chief could interpret this as a sign that the ancestors were satisfied with his own sacrifices. Moreover, if the plants grew it was because the adverse conditions had subsided. He could then predict, as Mutsora evidently did, that a good crop would be produced with the next planting. Whether or not a chief's interpretation was accepted appears to have depended on the degree to which he had previously gained the support of local ritual leaders, for without this support no chief could claim to control the land.

Finally, all these activities—tearing down huts, waiting for the plants to mature in the mounds, replanting the seeds, and waiting for them to germinate—took time, and time was the key to any *mwami's* success. For the longer he could keep the people believing that he had the situation under control, the more chance there was that the weather would change and that he would be credited with having ended the famine. Of course, the longer the crisis lasted, the more his credibility would be tested, and, if he did not have widespread support among local ritual leaders, his credibility might collapse, as Kivoto's evidently did in northern Isale and Ghotya's did in Vusekuli.

This description is, of course, an external view of events, which implies that Mutsora and other chiefs manipulated Bashu beliefs to achieve their own political ends. While this is possible, it need not be the case. The assertion that a rival chief was responsible for bringing on the famine was, after all, consistent with Bashu ideas concerning the need for ritual coordination and may well have been made in good faith. Moreover, I suggest that Mutsora, like all chiefs, had to possess a deep-seated faith in his own ritual power in order to convince people that he was indeed in control of events.

While Mutsora's success in maintaining his credibility during the period of famine can be attributed to the support he received from Kahindolya, and presumably from other groups, and to his status as the firstborn son of Luvango, an additional factor may have been the divination provided by the Bashu of Mbulamasi, who served as 'diviners of the land' for the people of Vusekuli. The Bashu of Mbulamasi claim that the people of Vusekuli came to their ancestor Vitsevo during the course of the famine to find out what was causing it. Vitsevo is said to have taken several bunches of unripened bananas and dropped them one at a time into a deep hole that was dug in his divining hut. The hole or shaft is said to have passed through the earth and thus through the abode of the ancestors and opened out into a pool located below the hill on which Vitsevo's divining hut was located. Before dropping each bunch of bananas into the hole Vitsevo proposed an explanation for the famine. He then released the bananas and the people from Vusekuli ran down the hill to the pool to see if the bananas had ripened during their descent. If they had, it was a sign that the ancestors were satisfied with the explanation and that plenty would return once the source of discontent had been removed. If the objects were unchanged, the explanation was deemed to be wrong.[31]

It would appear that Vitsevo's divination favored Mutsora's explanation, since Mutsora was eventually credited with ending the famine. If so, Vitsevo may have been influenced by the fact that both he and Mutsora were allies of Kivoto and that in assisting Mutsora he would be indirectly strengthening Kivoto's position in Isale. It is also possible that Vitsevo was allied directly to Mutsora, though there is no evidence of this.

On the other hand, divinations are frequently ambiguous, indicating the general source of misfortune without specifying the individuals involved. This was especially true in divinations for the land, presumably because of the potential danger of indicting a chief who might later prove successful in gaining power. Thus a divination for famine might indicate that a malevolent *mwami* had disrupted the land without specifying who the individual was. The divination then had to be interpreted by the people in accordance with their evaluation of the ritual credibility of existing chiefs. In other words, divinations did not necessarily provide a solution to the conflict between rival chiefs, but simply legitimized popular decisions. Vitsevo's divination therefore may not have directly aided Mutsora's cause.

In any case, Mutsora was credited with having ended the famine and was able to establish his ritual authority over the region that Ghotya had formerly controlled. He refrained, however, from driving Ghotya out of the region. Instead, he incorporated him into his system of ritual control by continuing the practice of participating together in the performance of land rituals. In this way he was able to add Ghotya's ritual authority to his own and thus increase his own ritual power.[32]

FAMINE AND SECESSION IN THE MUGHULUNGU AREA

The succession dispute between Kivoto and Muhoyo, combined with Mutsora's preoccupation with establishing his authority in Vusekuli, provided their cousin Molero, son of Visalu, with the opportunity to gain

TABLE 6
THE ALLIES OF MUTSORA AND MOLERO

Mutsora's Allies	Status	Office (within Kivoto's chiefship)
Bakira of Kirungwe	?	?
Kahindolya (Vutungu)	rainchief, successor to oldest chiefship in Isale	royal councillors (*baghani*)
Bito of Biabwe	healers of the land	owner of ritual seeds (*mikene*)
Basumba of Lukanga	first occupants of the land	keepers of royal fire; provided *musumbakali* for Vyogho's accession
Molero's Allies		
Batangi of Lulinda	royal ritualists for early Babito chiefs	keepers of royal regalia (*baghula*); *semombo*
Basumba of Makungwe	first occupants of the land	keepers of royal fire; provided skin of flying squirrel for *mbita*
Kisere s/o Mukirivuli	rainchief	*musingya*, presented *mbita*
Descendants of Kaherataba (Mughulungu)	clearers of the forest	royal buriers (*bakaka*)

control of the Mount Mughulungu area of Isale and to convince the people of this area that he was the legitimate successor to the early Babito chiefs of Isale, Mukunyu and Kavango.

Local traditions attribute Molero's success in carving out a section of Luvango's former domain for his own chiefship to his having ended the extended rains that hit the region after Luvango's death. Molero's ability to win credit for ending the rains can in turn be related to his success in building alliances with ritually influential groups in the area and in acquiring legitimacy in the eyes of the people of Mughulungu.[33]

Molero was raised by his maternal relatives, the Batangi of Lulinda, and, according to some informants, did not accompany his father into exile. The early support of the Batangi undoubtedly aided Molero in his attempt to establish himself as the legitimate successor to the early Babito chiefs of Isale, for the Batangi had directed the performance of royal rituals for the Babito since Busongora and had been responsible for transmitting and guarding the knowledge and regalia associated with Babito chiefship. They were also the guardians of Mukunyu's jawbone. The Batangi of Lulinda had refused to participate in the investitures of

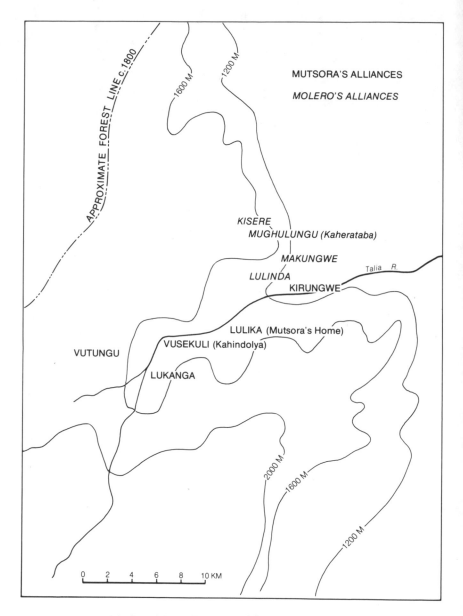

MAP 5. The Alliances of Mutsora and Molero

Luvango or his descendants, thus depriving them of an important source of legitimacy.[34]

When Molero grew older, he increased his support by forming several additional alliances. The first was with the descendants of Kaherataba, the family of 'clearers of the forest' who had welcomed Kavango and given him land at Kivika.[35] A second alliance was formed with the Basumba of Makungwe, who had been Kavango's allies and had been charged with relighting the royal fires following the investitures of Mukunyu and Visalu and with providing the flying squirrel skin used to fabricate their *mbita*. Like the Batangi of Lulinda, these ritually important groups had refused to recognize the authority of Luvango or his descendants.[36]

Molero's third alliance was with the senior branch of the former rain-chief Mukirivuli. As noted in chapter four, Mukirivuli had been Kavango and Mukunyu's patron and had provided Mukunyu with his *mbita*. Following his death, his two sons, Mbopi and Kisere, had fought over the succession, dividing the family's authority. Luvango had participated in this conflict in order to gain Mbopi's support. In reaction to this alliance, Kisere had thrown his support behind the losing cause of Visalu and subsequently participated in Molero's investiture.[37]

By tying himself to these groups, Molero reconstructed the early network of alliances created by his ancestors and in so doing helped to establish his claim to being the legitimate successor to the early Babito chiefs of Isale. His success in establishing this claim, and the important role that this success is seen as playing in ending the famine that occurred after Luvango's death, is reflected in the following tradition, which describes the events leading up to the establishment of Molero's chiefship:

> Mukunyu had two sons, Luvango and Visalu. At the death of Muku-nyu, Luvango and Visalu fought over the *mbita*. Luvango said that he was the firstborn son and that he should have the power. He told Visalu that he was capable of killing him if he disputed his authority. Visalu fled to Vungirwe among his mother's brothers [the Banisanza of Lisasa]. Later there was a great famine. A man named Lwasonga sought out Visalu and brought him back. Visalu made the sacrifice, the crops grew, and the people gave him the *mbita*.[38]

The reference to Visalu's returning to Isale in this tradition is contradicted by other traditions from the region, which state that Visalu died in exile. These traditions are supported by the location of Visalu's grave at Vungirwe. The reference to Visalu's having returned to Isale may therefore be meant to signify that Visalu's "authority" returned to Isale, while the reference to Lwasonga's having gone to find Visalu evidently refers to the fact that he went to get Visalu's jawbone, within which Visalu's authority rested. This interpretation is supported by the fact that Lwasonga's descendants presently guard Visalu's jawbone.[39]

The tradition, therefore, describes the renewal of Visalu's chiefship and how this renewal ended the famine in Isale. It thus attests to Molero's success in reconstructing his predecessor's chiefship and to the importance of this success for ending the famine and for establishing Molero's authority in Isale.

More specifically, local traditions suggest that having reconstructed the network of alliances with important ritual leaders in the Mughulungu region, which his father and grandfather had created, Molero was able to establish his own claim to ritual legitimacy and then, during the famine, to reinforce this claim by attributing the famine to Luvango's usurpation of his family's authority. The acceptance of this explanation by the people of Mughulungu resulted in Molero's investiture, which is said to have coincided with the return of drier weather.

CONCLUSION

The years following Luvango's death saw the fragmentation of Babito authority in Isale and the creation of four separate spheres of political influence. While this development resembles political processes that occurred elsewhere in eastern and southern Africa during the precolonial period, the specific form that it took among the Bashu reflected Bashu ideas about chiefship and the need to create spheres of ritual influence.

In each political arena rivals for chiefship attempted to establish their legitimacy as chiefs by gaining the support of local ritual leaders, thereby establishing control over the performance of land rituals. The relative success of each competitor was then tested and defined during the extended period of rains that hit Isale several years after Luvango's death. Success in maintaining legitimacy as a chief in the face of this natural disaster was directly related to a competitor's prior success in building a sphere of ritual control and in establishing ritual credibility. For this success permitted him to present and maintain an explanation for the disaster that deflected ultimate responsibility for its occurrence onto a rival, and thus to maintain his legitimacy. From a popular perspective, of course, it was not a question of gaining credibility but of actually establishing ritual control over the land. A chief who established this control had the power to regulate the well-being of the land, while a chief who failed to achieve control could not regulate the land and was seen as a threat to its well-being.

The ability of three separate competitors to provide explanations for the famine and maintain their explanations until the crisis had passed indicates, in part, the localized nature of social relations among the Bashu at this time. Yet it also indicates that this particular famine, serious though it was, did not last so long as to lead the people of any given region to look beyond the region for an explanation of the famine. It did not, therefore, lead to further confrontations and testings of authority between leaders of adjacent areas of Isale or to a wider consolidation of ritual authority.

Local ideas about the nature and function of chiefship appear to have had a similar impact on subsequent relations between various branches of the Babito, as well as on the overall position of the Babito vis-á-vis non-Babito chiefs in Isale. To begin with, while Muhoyo, Kivoto, Mutsora, and Molero competed for control of various areas of Isale, they subsequently cooperated in the performance of land rituals. Thus it is said that while they fought by day they sacrificed together by night. In other words, political divisions did not break the ritual community that Luvango had initiated. This does not mean that each branch did not continue to compete for the support of local ritual leaders and attempt to expand their respective spheres of ritual, and thus political, influence at the expense of other branches, but rather that overall ritual cooperation superseded these shifts in each branch's sphere of influence. This somewhat paradoxical pattern was still observable during the course of my research, when, despite continued feuding over who should control Isale, all the Babito chiefs in Isale participated in the performance of a sacrifice that from a Bashu viewpoint prevented a plague of locusts from descending upon Isale.

I suggest that this pattern of ritual cooperation reflects the importance that the Babito attached to maintaining a single sphere of ritual influence over Isale. The maintenance of this unified sphere of ritual cooperation had political advantages, for it provided the Babito with a greater claim to ritual and political legitimacy than any of their non-Babito rivals. Specifically, it meant that each Babito chief could draw on ritual resources that exceeded the limits of his personal alliances when competing with non-Babito rivals. In cases where the relationship between two branches of the family was particularly close, as was apparently the case between Kivoto and Mutsora, this ability to draw on outside resources may have involved direct assistance, as suggested by the possibility that

Mutsora acquired the assistance of Kivoto's allies, the Bashu of Mbulamasi. Yet, even where relations were more distant, a chief's credibility may well have been heightened by his status as a Mubito and his connection to the wider field of ritual cooperation that existed among the Babito. It was this cooperation, as much as their superior economic resources, that, I suggest, prevented the previous chiefs of Isale from taking advantage of the political divisions and famine that followed Luvango's death in order to reassert their own position as chiefs. The Babito, therefore, were able to maintain their overall domination of Isale, despite divisions within their ranks and continued attempts by each branch of the family to claim a larger share of the region as their own.

It is perhaps significant that this wider sphere of ritual cooperation among the Babito of Isale did not include the Babito chiefs of Malio and Maseki. This exclusion may reflect the continued importance of ecological divisions within the Bashu region and the obstacle that the divisions created for ritual cooperation. These divisions may also account for why the chiefs of Isale made no effort to expand their political influence into Maseki or Malio, despite political "overcrowding" in Isale. On the other hand, it is possible that some sort of informal or formal agreement existed between the chiefs of the three areas concerning respective spheres of influence, or that the chiefs of Malio and Maseki were simply too strong to permit the chiefs of Isale to expand their influence beyond Isale.

Nonetheless, the history of political relations in Isale following Luvango's death can in general be seen to reflect Bashu ideas about the nature and function of chiefship and politics. During the last years of the nineteenth century and the early years of the twentieth, these ideas were challenged by the introduction of new economic and political resources and by the establishment of Belgian colonial rule. The impact of these changes on Bashu political ideas and on the nature of Bashu chiefship will be examined in the next three chapters.

CHAPTER SIX

TRADE, FIREARMS, AND THE POLITICS
OF ARMED CONFRONTATION
IN THE UPPER SEMLIKI VALLEY REGION

The second half of the nineteenth century saw the growth and expansion of long-distance trade in slaves and ivory and the introduction of new economic resources into the East African interior. Previous studies have described how the growth of this trade stimulated major changes in the social, economic, and political environments of the societies it touched (Hartwig, 1976; Feierman, 1974; Shorter, 1972; Gray and Birmingham, 1970; Alpers, 1969). Politically, it has been argued that the growth of trade contributed to a general secularization of political authority, with military force and the control of trade replacing or supplementing religion as the primary basis of authority (Roberts, 1969). In some cases, this transformation was initiated by existing chiefs or kings, such as Mutesa of Buganda, Kabarega of Bunyoro, and Rwabugiri in Rwanda, who employed firearms and wealth from trade to bolster their political authority and decrease their need for ritual legitimacy. In other cases, the change was brought about by "new men," such as Mirambo among the Nyamwezi and Semboja among the Shambaa, who lacked traditional sources of legitimacy but were able to gain control of firearms and trade and use these resources to overthrow traditional rulers.

While there can be little doubt that the expansion of trade and firearms altered the political environment of eastern Africa, the shift from religious to secular authority was by no means a necessary outcome of this transformation. For example, Semboja, despite his access to trade and firearms, chose to maintain the sacral character of Shambaa kingship, placing his son as king at the royal capital of Vugha while maintaining his own power base near the plains.[1] So too, Bashu ideas about the nature and function of chiefship exhibited a remarkable persistence and were, in fact, reaffirmed by changes brought about by the expansion of trade and firearms into the upper Semliki Valley at the end of the nineteenth century. In the present chapter, I will examine this expansion and how it affected the political development of the upper Semliki Valley region as a whole. In the next chapter, I will describe its specific impact on politics in Isale.

TRADE AND VIOLENCE
IN THE UPPER SEMLIKI VALLEY

The growing role of trade as a primary basis of political power in eastern Africa made the control of trade items and routes a major focus for political competition during the second half of the nineteenth century. By the 1880s this competition had spread into the upper Semliki Valley, creating new opportunities for political advancement for the peoples of the region.

The upper Semliki Valley was economically important for several reasons. First, as we have seen in earlier chapters, it was the site of a major salt deposit at Katwe and thus the center of a thriving salt trade that stretched over wide areas of western Uganda, Rwanda, and eastern Zaire. Second, the valley and neighboring mountain and forest regions contained large herds of elephants, which made them important sources of ivory. The demand for this ivory grew steadily during the last decades of the nineteenth century as herds of elephants closer to the coast were depleted.[2] Third, the valley provided a corridor through which traders could move with relative ease from the ivory-rich forest regions to the north and west of the valley to Busongora, Nkore, and the ivory markets of East Africa. The valley thus became a major route in the ivory trade. Finally, the valley contained many cattle, which, while less important than ivory or salt as items of trade, were nonetheless an important economic and political asset within the kingdoms of western Uganda.

The economic resources of the Semliki Valley attracted several outside groups during the 1880s and 1890s. These included the armies of the Nyoro king Kabarega; ivory traders from the Manyema region of eastern Zaire, under the leadership of Kilongalonga; and political refugees from the expansionist wars of the Rwandan king Rwabugiri, under the leadership of Karakwenzi. To this mélange of competing armies was added King Leopold's colonial agents, who established posts in the valley in 1896 and 1897, in conjunction with the Belgian Nile Expedition and in order to prevent the westward expansion of British influence from Uganda. While these groups were motivated by somewhat different interests, they competed with one another for control of the valley and its resources, using firearms and terror tactics to achieve their goals. These tactics, the politics of armed confrontation, were subsequently borrowed by local competitors for chiefship in the mountains surrounding the valley, including Isale. The intrusion of foreign groups into the upper Semliki Valley consequently affected the political development of these neighboring regions.

Kabarega came to the throne of Bunyoro in 1871 and quickly set about reestablishing Nyoro control over the regions that had become independent of Bunyoro during his predecessors' reigns. To accomplish this, Kabarega organized a standing army in Bunyoro and supplied many of his soldiers with firearms acquired through trade with Sudanese ivory traders. As a result of these changes, the Nyoro army became a powerful military force that could be put into the field at short notice and was generally better trained and armed than the temporary provincial levies of Buganda, Toro, and Nkore. Kabarega employed his army to plunder and terrorize his opponents' subjects and to undermine his opponents' political credibility and support. While armed confrontation was a common tactic in the interlacustrine region, Kabarega employed it more frequently and with greater effectiveness than any of his rivals or predecessors (Ingham, 1975: 55; Uzoigwe, 1970: 11–12).

By the early 1880s, Kabarega's armies had defeated the princes of Toro, whose predecessors had seceded from Bunyoro around 1830. He then began the reconquest of Busongora. Kabarega appears to have had several reasons for reconquering this region. First, he wished to reestablish Nyoro control over the salt lakes at Katwe and Kasenyi and thus over the extensive salt trade that centered on these lakes. Second, he was attracted by the large herds of cattle in Busongora (Ingham, 1975: 58; Tosh, 1970: 105). Finally, he may have wished to secure a foothold on the southern trade route, the northern route having dried up following the Mahdist takeover in Khartoum in 1885 (J. B. Webster, personal communication).

By 1889 a Nyoro army under the command of Ireeta, a Munyankore general who had allied himself with Kabarega, had occupied Katwe, gaining control of the salt lake and channeling the flow of salt northwards to Bunyoro (Ingham, 1975: 57–58; Stanley, 1890: 509–10). While Ireeta was evidently content to limit his activities to the Katwe region, his successor, Rukara, expanded Nyoro influence to the west and north, raiding extensively in the plains, where his soldiers captured large numbers of cattle, and in the mountains, where they plundered Nande/Konjo settlements. There is no evidence, however, that Rukara sought to establish permanent political control west of the Semliki or along the western slopes of the Ruwenzoris; instead he chose to use these areas as convenient reservoirs for food, livestock, and women. The Nyoro army dominated the area around Katwe until Lugard's arrival in 1891. With the assistance of local salt traders, Lugard's forces drove Rukara away from Katwe and established a fort in the area. The Nyoro army returned briefly in 1893 but was later recalled by Kabarega to assist in his final battle against British-led forces (Stanley, 1890: 284; Ingham, 1975: 58).[3]

A similar though somewhat varied pattern of activity was initiated by Manyema ivory traders, who were also attracted to the Semliki Valley during the 1880s. Armed bands of Manyema operated throughout the mountains and forests of what is now the Kivu region of Zaire during the 1870s, 1880s, and 1890s, trading for, or extorting, ivory from local populations. The Manyema slowly expanded their activities northwards and by 1891 had established themselves at Miala in the upper Semliki Valley, to the northwest of Katwe, and in the lower Semliki Valley, where their leader, Kilongalonga, had his main camp. From their camps in the region, the Manyema raided into the Ituri, the Semliki Valley, the Mitumbas, and the northern slopes of the Ruwenzoris in search of ivory (Joset, 1939: 11–12; Lugard, 1959: 353). A second group of Manyema, under the leadership of Lukundula, operated among the southern Nande as far north as the Baswaga chiefdoms of Bulengya and Bukenye.

The Manyema, like Kabarega's army, possessed large numbers of firearms, which they employed to extract ivory from the local population of the forested areas to the north and west of Isale, claiming one tusk for every elephant killed in the regions immediately adjacent to their camps and levying tribute in ivory from more distant villages at irregular intervals (Joset, 1939: 11; Lugard, 1959: 334). Resistance to Manyema demands was met with massive retaliation, as witnessed by the people of the Baswaga chiefdom of Mwenye, who attacked a Manyema collection party, killing some thirty men. The Manyema returned with an estimated three hundred rifles and ravaged the northern half of the chiefdom, destroying villages, killing men, and enslaving women and children (Bergmans, 1970: 51–52).

Among the Bashu, the Manyema made no demands for ivory but raided only for food, livestock, and occasionally women. This apparent lack of interest in ivory is curious, given Bashu involvement in the ivory trade at this time. The Bashu killed elephants locally and traded their ivory for cloth, beads, iron hoes, and, most importantly, goats and some cattle, brought by Bahima traders, known locally as *batembesya*, from the Semliki Valley.[4] While the volume of this trade was not as extensive as that emanating from the forest regions to the north and west of Isale, it was, nonetheless, significant enough to cause Bashu chiefs to impose a tax on the trade of one tusk for every elephant killed, to prevent commoners from acquiring too much wealth.[5]

It is possible that the Manyema saw the Bashu region primarily as a source of agricultural produce and livestock and concentrated their ivory collection activities in areas where ivory was more plentiful. The Bashu practice of burying their ivory until the *batembesya* arrived may have encouraged this pattern. On the other hand, it is interesting to note that, according to one informant, the Manyema eventually acquired much of the Bashu ivory through trade with the *batembesya,* and that in exchange for this ivory they gave the *batembesya* livestock, a large portion of which they had presumably confiscated from the Bashu.[6] In other words, a portion of the livestock that the Manyema used to acquire Bashu ivory from the *batembesya* was eventually regained by the Manyema through raids on the Bashu region. This suggests that the ivory trade involving the Bashu, *batembesya,* and Manyema may in reality have been a system of ivory extraction involving trade, raiding, and the recirculation of live-stock.

Such a system, if it in fact existed, would have been particularly efficient from a Manyema viewpoint, for it would have insured a maximum supply of ivory at a minimal cost. The system would have encouraged the Bashu to acquire and trade as much ivory as possible by creating an illusion of profitability. This illusion would be maintained by the triangular nature of the system, which allowed the groups trading for ivory and the groups raiding for food and livestock to appear to be independent of one another. This would have led the Bashu to believe that setbacks suffered from raids were momentary and random in nature, and that their loss of livestock in a Manyema raid could be recouped through subsequent exchanges in ivory with the *batembesya.* In reality, however, the losses were an inherent and regular part of the system. The costs of administering such a system from a Manyema standpoint would have been minimal. It would have required only a small outlay of livestock, lost to *batembesya* profit taking or Bashu consumption, and the cost of periodic raiding, which, given the possession of firearms, could be carried out by relatively small bands of men, freeing the main Manyema forces to concentrate their activities on more productive ivory areas. Had the Manyema attempted directly to extort ivory from the Bashu, it would undoubtedly have required a greater investment of manpower and discouraged the Bashu from hunting. It is possible, moreover, that this increased investment may not have been worth the relatively limited amounts of ivory that could be extracted from the region. The combined trading-raiding system may thus have provided an efficient means of extracting ivory from a region in which the supply of ivory was restricted. Obviously, further research needs to be carried out on Manyema activities in the region before we can determine the extent to which this system actually operated.

The third group to be attracted to the upper Semliki Valley at this time was led by Karakwenzi. According to Joset (1939: 28), Karakwenzi was an important Tutsi chief from Lake Nyabirongo in Rwanda. However, Jim Freedman's research in Ndorwa, located on the present border between northern Rwanda and the Kigezi district of Uganda, indicates that Karakwenzi came from Ndorwa, where he was a famous medium for the spirit of Nyavingi, the legendary queen of Ndorwa who became the center of a powerful religious cult in the western lakes region. Karakwenzi's association with Nyavingi is supported by his subsequent use of Nyavingi to legitimize his political position in the upper Semliki Valley. Karakwenzi evidently left Ndorwa around 1880 in response to King Rwabugiri of Rwanda's military incursions into Ndorwa. He and his followers fled northward, traversing Ufumbiro and Rutshuru and passing through Nkore. He arrived at Katwe sometime between the Nyoro invasion of the early 1880s and Stanley's arrival in 1889.[7]

Unlike the Banyoro and Manyema, who were predatory raiders, Karakwenzi attempted to establish an organized political system in the upper Semliki Valley, with a hierarchy of subchiefs and a regular system of tribute payments. His territorial organization may have been based on a Rwandan model, for, like Rwanda, Karakwenzi's area of control was divided into districts, each with a capital under the leadership of a "queen" (Joset, 1939: 28).

The extent to which Karakwenzi employed firearms and terror tactics to establish his chiefship is unclear. Data from the Maseki region of the Mitumba Mountains, where Karakwenzi settled in the mid-1890s, suggest that he employed considerable force to establish his authority in the region. In addition, Joset (1939: 28) claims that Karakwenzi raided throughout the Semliki Valley and neighboring regions, collecting ivory that he exported to East Africa. He is also said to have employed firearms obtained from the British ivory trader Charles Stokes. Karakwenzi thus appears to have used tactics similar to those employed by the Manyema and Banyoro.

On the other hand, Stanley's references to Karakwenzi (1890: 334) indicate that Karakwenzi was viewed as something of a savior by the people of the region, whom he protected from Kabarega's army and the Manyema. Stuhlman (1894: 278) who, along with Emin Pasha, met Karakwenzi on their journey through the Semliki Valley, had a similar impression of Karakwenzi's relationship to the local population. Even Joset (1939: 29) notes that the local population gave Karakwenzi guns taken from Manyema they had killed. He also claims that Karakwenzi made alliances with a number of local chiefs, including the *mwami* Tsombira (Nzumbia in Belgian records) of the Banisanza of Lisasa. These statements appear to contradict the Maseki traditions as well as Joset's own evaluation of Karakwenzi.

The scarcity of historical traditions from the upper Semliki Valley, which is now a national park, makes it difficult to evaluate these conflicting claims. It is possible, however, that Karakwenzi viewed the people living in the region just north of Lake Edward (which formed the core of his chiefdom and was the region from which Stanley and Stuhlman collected most of their data on Karakwenzi) as his subjects who qualified for his protection, whereas he viewed the people living in more distant regions, and especially in the mountains, as strangers subject to attack.

Karakwenzi appears to have legitimized his chiefship by claiming to be a descendant of Nyavingi. Thus, one of Stanley's porters from Katwe claimed that Karakwenzi was a "true son of the Wanyavingi" (1890: 334). Bashu traditions state that Karakwenzi built a number of shrines to Nyavingi along the north shore of Lake Edward. He may, in fact, have introduced the Nyavingi cult to this region, though it seems more likely that Nyavingi was already known north of the lake through trading contacts to the south and migrations from Rwanda and Ndorwa, and that Karakwenzi simply employed his own Rwandan or Ndorwan origins and military power to establish his claim to being the son of Nyavingi.[8]

Karakwenzi's activities brought him into conflict with both Kabarega's soldiers and the Manyema, for, by organizing the population of the upper Semliki Valley, he impeded the raiding activities of these groups. In addition, his participation in the ivory trade placed him in direct economic competition with the Manyema. His initial conflict with these groups, however, stemmed from his unsuccessful attempt to control the salt lake at Katwe. This led to an encounter with Kabarega's men, which, in turn, brought him into contact with the Manyema.

To gain control of Katwe, Karakwenzi allied himself with Kakule, the chief of the Bakingwe salt traders. This alliance allowed him to drive the Banyoro army away from Katwe. Kabarega's men returned, however, with a greater force and drove Karakwenzi and his men across the Semliki, where they were stationed when Lugard arrived in 1891 (Stanley, 1890: 284, 334; Lugard, 1959: 250–52). Following his defeat, Karakwenzi attempted to enlist Manyema support to regain control of Katwe. The Manyema, however, had no interest in becoming involved in Karakwenzi's war and instead took him prisoner and held him for ransom. Following his release Karakwenzi attempted unsuccessfully to avenge himself by driving the Manyema out of the upper Semliki Valley. Lugard (1959: 268) reported in 1891 that Karakwenzi was at war with the Manyema and that all of the villages near the border between Karakwenzi's settlements and those controlled by the Manyema were deserted. Karakwenzi, unable to dislodge the Manyema and prevented from expanding his authority to the east by Kabarega and later the British, moved into the Maseki region of the Mitumbas around 1894.[9]

While Karakwenzi was settled in Maseki, the Belgians established their first posts in the upper Semliki Valley at Katwe, several miles west of the salt lake on the west bank of the Nyamgassa River, and at Kirimi, located on the southern slopes of the Ruwenzoris. At the end of 1897 a third post was established at Beni, on the west bank of the Semliki, just south of the equatorial forest. From these posts, the Belgians extended their political authority over the upper Semliki Valley. The posts later served as staging areas for the occupation of the surrounding mountain regions (Joset, 1939: 11).

The Belgian presence in the valley threatened the economic and political interests of Manala, the local Manyema leader, and Karakwenzi. Both men, therefore, attempted to drive the Belgians out of the area. The Manyema attacked the Belgian post at Kirimi shortly after its construction. However, the post commandant, Lt. Sannaes, was warned of the attack by the Nande chief Mbene and was able to withdraw his small detachment of men to Katwe before the attack. But the Manyema success in driving the Belgian force from Kirimi was short-lived; soon after their attack, a Belgian force under the command of Lothaire and Henry succeeded in dislodging the Manyema from their main centers of activity in the forest to the north of Beni. The Manyema forces retreated south through the valley and attacked the Belgian post at Katwe on April 28, 1897. This time the attack was repulsed with the help of British forces located east of the Nyamagassa. The Manyema suffered heavy losses and retreated westward into the forest, terminating their activities in the valley.[10]

In June of 1897, Karakwenzi, who had abandoned his mountain settlement and returned to the plains, attacked and destroyed the Belgian post at Kirimi in an apparent attempt to reestablish his control of the upper Semliki. The Belgian force of sixty men withdrew to Katwe, against which Karakwenzi launched a second attack, forcing the Belgians to retreat to Fort Portal in Uganda. The Belgians returned several months later and built a new post at Kasindi, near the present Zaire-Uganda border crossing. The following year they succeeded in arresting Karakwenzi and exiling him to Mawambi in the Ituri (Joset, 1939: 12).

By the end of 1898, therefore, the Belgians had succeeded in defeating their major rivals for control of the upper Semliki Valley. They then turned their attention to establishing their control over the mountain regions bordering the valley. To accomplish this, the commandant of Beni enlisted the support of the Nande chief Mbene, who had earlier demonstrated his support by warning Lt. Sannaes of the impending Manyema raid on Kirimi, and Mbene's subchief Kirongotsi. The Belgians provided these chiefs with firearms and encouraged them to pacify the surrounding areas. Both men took advantage of this Belgian patronage and raided extensively throughout the region surrounding Beni (Joset, 1939: 14). Kirongotsi's activities extended into northern Isale, where he is remembered as a cruel man who stole food, women, and livestock.[11]

THE CHANGING POLITICAL ENVIRONMENT OF THE MITUMBA MOUNTAINS

Political competition and warfare in the upper Semliki Valley had a major impact on the people who inhabited the mountains adjacent to the valley, and especially on the peoples of Isale, which, because of its topography, offered plains groups easy access to the mountains. The repeated raids of the various groups operating in the valley caused great hardships and loss of life among the peoples living in the mountains. The passage through northern Isale of Batetela recruits who had mutinied against the Belgian commander Dhani in October of 1897 only added to this suffering. One informant describing life in Isale during this period stated, "We were like animals and these men were the hunters."[12] The sufferings produced by these raids were not, however, the only consequence of expatriot activities in the upper Semliki Valley. The expansion of the ivory trade, with the introduction of firearms and terror tactics, provided local competitors for chiefship with new resources and opportunities for political advancement.

The ivory trade, despite its inherent inequalities, increased the political authority of mountain chiefs. Chiefs were less susceptible than commoners to losses suffered through raids, even though their superior access to ivory provided them with large numbers of cattle and goats. This was because chiefs seldom kept the livestock they acquired through trade for any length of time. Instead, they quickly redistributed them to their supporters as gifts. By doing so, they converted livestock, which was a vulnerable asset, into political capital in the form of social debts, which were intangible and thus safe from raids. In this way, they were able to use the trade to reinforce their ties with local ritual leaders and lineage heads.

In addition, while the repetitive raids by plains groups challenged a chief's ability to insure the well-being of his subjects, they also worked, at least initially, to strengthen his hold over his subjects. For it was to his chief that a man who had lost his livestock would look for assistance. The raids therefore increased commoner dependence on chiefly beneficence and further solidified the chief's support base. Ultimately, however, the raiding appears to have eliminated all livestock in the mountains, ending any marginal benefit that the chiefs might have gained from it.

As elsewhere in eastern Africa, the introduction of firearms enabled some chiefs to reinforce their authority and decrease their dependence on local ritual leaders. At the same time, it allowed for the emergence of "new men" who, for one reason or another, lacked traditional legitimacy and found in the acquisition of firearms an alternative path to political power.

To the north of Isale, Tsombira, the chief of the Banisanza of Lisasa, was able to use firearms to expand and reinforce his political authority in the region. Tsombira's home was located near the plains on the lower slopes of Mount Lisasa. He was, consequently, one of the first Nande chiefs to come into contact with the warring groups that dominated the upper Semliki at the end of the nineteenth century. Tsombira's people suffered heavy losses from several raids by the Manyema traders stationed at Miala. It was perhaps for this reason that Tsombira welcomed Karakwenzi and became his ally. This alliance provided Tsombira and his people with protection from further Manyema raids as well as a small number of firearms that Tsombira received as a gift from Karakwenzi. In return, Tsombira agreed to recognize Karakwenzi's overlordship and to support him in his war against the Manyema.[13]

Tsombira subsequently employed the firearms received from Karakwenzi to expand his own political authority in the mountains surrounding Lisasa and to raid more distant areas, including northern Isale. Tsombira's military superiority is reflected in the following tale told about him by people in northern Isale:

> Tsombira had the power to turn himself into an elephant and travel through the country ravaging people's gardens. When the people saw the elephant they tried to drive it away by hurling their spears at it. The elephant would then move off, the spears lodged in its side. Later Tsombira called the people to come to his home. When the people arrived they were amazed to find all of their spears in a circle around Tsombira's *kyaghanda* [meeting house].[14]

To the southwest of Isale, in the Baswaga chiefdom of Bukenye, the chief Mayi provides a more extreme example of the use of firearms to achieve political authority. Mayi's claim to chiefship prior to his acquisition of firearms was more tenuous than Tsombira's. Mayi had been appointed as regent for the young prince Kambere Bwabu. Mayi soon took advantage of the situation and established himself as chief in Bukenye. To accomplish this, he allied himself with Manyema traders attached to Lukundula, whose headquarters were located in Walikale to the south of the Nande region. While the exact nature of Mayi's alliance with the Manyema is unclear, it evidently resulted in Mayi's acquisition of a number of firearms. With these rifles he organized a small army of men known locally as *basombi*. Some of these warriors are said to have been attracted from neighboring regions and appear to have been unattached men dislodged from their homes by Manyema raids. The *basombi* thus resemble the *ruga ruga* irregulars of Mirambo and Nyungu-ya-Mawe in Tanzania. Mayi employed the *basombi* to terrorize the people of Bukenye, raiding the homes of recalcitrant groups, killing their leaders and taking whatever wealth they found. Mayi is said to have hanged many of his victims as a symbol of his power and a warning to those who might try to oppose him. The people of Bukenye claim that nearly every hill in the chiefdom was ornamented with Mayi's victims. He is also said to have constructed a hut at his home, in which he kept the skeletons of his victims to intimidate visitors. Through the use of these terror tactics Mayi established his claim to chiefship in Bukenye.[15]

The Nande chief Mbene—who, as we have seen, eventually became an ally of the Belgians—is a second example of a "new man" who used firearms and the politics of armed confrontation to establish his political authority. Mbene is said to have been a blacksmith from the Baswaga chiefdom of Buyora, who migrated to the area near the present town of

Beni during a famine. Mbene became a client of the area's chief, Mohanga, and served as Mohanga's representative to the first Europeans to arrive in the area. Mbene subsequently capitalized on his European contacts and acquired European recognition as chief of the region. The first European with whom Mbene made contact, either on his own or as Mohanga's representative, was the ivory and firearms dealer Charles Stokes, who operated in the upper Semliki Valley region during the early 1890s. Mbene's village subsequently became a depot for Stokes's trading supplies and a center for his activities in the upper Semliki Valley and surrounding regions. It was probably from Stokes that Mbene acquired his first firearms. With these guns he was able to assert his own claim to chiefship and eventually to usurp Mohanga's authority in the area.[16]

In 1895 Mbene's strategy of using European contacts to advance his political career suffered a temporary setback when Stokes was arrested and executed by the Belgian commander Dhani for illegal trafficking in firearms. Stokes's supplies and firearms at Mbene's were subsequently confiscated. Mbene was thus deprived of his European ally. However, he soon found a way to gain the confidence of this new group of Europeans when, in 1896, the Manyema leader Manala tried to enlist his support in an attack on Kirimi. As noted above, Mbene warned the Belgian commander of the attack and won his gratitude. In the following year, Mbene's assistance was repaid by the construction of a Belgian post at Mbene's village, which the Belgians called Beni. The Belgians subsequently recognized Mbene's authority in the region and provided him with new firearms.[17]

The histories of Tsombira, Mayi, and Mbene provide evidence of how changes in the economic and political environment of the upper Semliki Valley provided local competitors for chiefship in regions surrounding the valley with new opportunities for political advancement and, by doing so, affected political development in the Mitumba Mountains. It was within the context of these changes that political competition was carried out in Isale during the 1880s and 1890s.

CHAPTER SEVEN

"NEW MEN" VERSUS RITUAL CHIEFS IN ISALE

As elsewhere in the upper Semliki Valley region, the expansion of the ivory trade and the introduction of firearms in Isale affected patterns of political competition and development during the last two decades of the nineteenth century. Nonetheless, Bashu ideas about the ritual nature of chiefship continued to affect the choices and actions of competitors for chiefship and to provide the Bashu with a conceptual framework for explaining the events that occurred at this time. In the end, in fact, the events of the period reaffirmed Bashu acceptance of the need for ritual coordination and reinforced the politics of ritual chiefship.

In chapter five we saw how political competition following Luvango's death divided political authority in Isale among four separate branches of the Babito. With time, this balance of power was altered as the processes of support-building, subversion, and succession continued, increasing the authority of some chiefs while decreasing that of others. As a result of these shifts in authority, two men emerged as major rivals for political authority in Isale during the 1890s: Vyogho, the son of Kivoto, and Kasumbakali, the son of Muhoyo.

On the surface, Vyogho and Kasumbakali appear to have employed radically different strategies to achieve political power in Isale. While Vyogho used the tactics of alliance formation and peaceful subversion employed by his ancestors, Kasumbakali adopted the new tactics of armed confrontation based on the use of firearms and terror. Competition between these two leaders, therefore, appears to represent a confrontation between ritual and secular politics. A closer examination of each man's actions, however, reveals that they both sought to obtain the support of local ritual leaders and former chiefs and to create spheres of ritual influence in Isale. They thus shared the same political goal and differed only in regard to the means they employed to obtain this goal.

VYOGHO AND THE POLITICS OF RITUAL CHIEFSHIP

Vyogho began his political career with the backing of several important groups whose support he had inherited from his father. As noted in

chapter five, the core of Kivoto's support prior to his unsuccessful encounter with Muhoyo consisted of the Bashu of Mbulamasi, the Bahera
of Ngukwe, and his brother Mutsora. Following his defeat by Muhoyo
and his departure from Mutendero, Kivoto strengthened his position by
forming marriage alliances with the Basyangwa of Muluka, who were
'clearers of the forest' in southern Isale, and the Batangi of Lulinda, who
had directed the performance of royal rituals for Kivoto's ancestors since
Kitara. Kivoto's alliance with the Batangi was particularly important, for
they had previously refused to recognize Kivoto's branch of the Babito,
insisting that Visalu and his descendants were the only legitimate chiefs
of Isale, and that Kivoto, Muhoyo, and their father, Luvango, were
merely pretenders. They had thus denied Luvango and Kivoto the legitimacy that their support could provide. The decision by the Batangi of
Lulinda to support Kivoto was apparently initiated by a succession dispute that followed the death of Visalu's son Molero, whom the Batangi
had invested and who was their *muhwa*, sister's son.

Molero is said to have designated his son Vukendo as his successor. This
choice was subsequently confirmed by the members of the *basingya*,
including the Batangi of Lulinda. However, Vukendo was still a young
man, and Molero's eldest brother, Kivere, was appointed as a regent until
such time as Vukendo was old enough to take on the responsibilities of
chiefship. Kivere was not content to remain regent for long and soon used
his position to win the support of some of Molero's former allies, including the Basumba of Makungwe and the descendants of the rainchief
Mukirivuli. Having gained the support of these groups, Kivere claimed
the chiefship for himself and forced Vukendo to flee to Vutengera,
where he was given land by Kivoto. Kivere was assassinated by Vukendo's supporters some years later. However, they were unable to bring
about Vukendo's return and Kivere's son, Kamabo, was invested as chief
over the Mughulungu area.[1]

The succession struggle between Kivere and Vukendo split the support base that Molero had constructed in the Mughulungu area, seriously
weakening the political position of this branch of the Babito. It was in
the wake of this split that the Batangi of Lulinda threw their support
behind Kivoto, who had given refuge to Vukendo. Batangi support
strengthened the claim of Kivoto and his descendants that they were the
legitimate successors to the first Babito chiefs of Isale.[2]

Following his investiture in the early 1890s, Vyogho set about improv-
ing the position he had inherited from his father. He began by increasing
the vested interest of the Batangi of Lulinda in his chiefship by making

them his family's *baghula*, guardians of the royal regalia, a position that had formerly been held by the Bashu of Mbulamasi.[3] Accordingly, Vyogho asked the descendants of Ghotya for the hill of Ihera, located in the Vusekuli area, as a place upon which a section of the Batangi of Lulinda could settle.

Vyogho's replacement of the Bashu of Mbulamasi as *baghula* was part of a long-term process involving the manipulation of alliances by Luvango's descendants in order to strengthen their political position in Isale by increasing the number of important groups with a vested interest in their chiefship. As noted in chapter four, it was Luvango who had first established an alliance with the Bashu of Mbulamasi by making them his family's *baghula*, and by taking his *mombo* from among them. During Kivoto's reign the Bashu of Mbulamasi retained the position of *baghula* but were made to share the position of *semombo*, 'giver of the *mombo*,' with the Bahera of Ngukwe. This sharing was accomplished by requiring the *mombo*, who came from Ngukwe, to pass through Mbulamasi on her way to Kivoto's home, and to 'sit on the knees' *(-tambika)* of the Bashu of Mbulamasi. This 'sitting on the knees' symbolized the Bashu of Mbulamasi's sociological paternity of the *mombo*. Finally, during Vyogho's reign, the Batangi of Lulinda replaced the Bashu of Mbulamasi as *baghula* and the position of *semombo* was divided among three groups: the Bashu of Mbulamasi, the Bahera of Ngukwe, and the Batangi of Lulinda. Thus, Vyogho's *mombo* passed through Mbulamasi and Lulinda on her way from Ngukwe to Vyogho's home, making all three groups sociological mother's brothers to Vyogho's successor.[4] Through these changes Kivoto and Vyogho reduced their dependence on the Bashu of Mbulamasi, while apparently maintaining this group's continued support. At the same time, their manipulation of the position of *semombo* maximized the number of supporters with a vested interest in their chiefship.

Vyogho further reinforced the support he had inherited from his father by strengthening his ties with the descendants of his paternal uncle Mutsora, who had been Kivoto's *mukulu*. The continued support of this group was important because of their control of Vusekuli, which, as we have seen, formed a political bridge between northern and southern Isale. To insure the continued support of Mutsora's descendants, Vyogho—or more likely his *basingya*—invested Mutsora's son Mukiritsa as Vyogho's *mukulu* rather than giving the position to Vyogho's eldest brother, who normally would have inherited it. This office gave Mukiritsa a vested interest in Vyogho's chiefship and provided him with considerable influence among Vyogho's followers. It is, of course, possible that, given the importance of his support, Mukiritsa was able to demand that the position remain in Mutsora's lineage.[5]

Vyogho expanded his inherited support base by forming three new alliances with ritually important groups. These alliances increased the geographical distribution of his sphere of ritual influence. Vyogho took his first wife from among the descendants of Kaherataba, who were the 'clearers of the forest' on Mughulungu. Kaherataba's family, like the Batangi of Lulinda, had been early allies of the Babito in Isale and had, following the succession dispute between Luvango and Visalu, supported Visalu's successor, Molero. During the succession dispute that followed Molero's death they had similarly refused to recognize Kiverc's unsurpation of Vukendo's authority. They do not appear, however, to have followed the Batangi in immediately recognizing Kivoto as the legitimate successor to the original Babito chiefs of Isale. For there is no evidence of an alliance between them and Kivoto. Vyogho evidently convinced them to form a marriage alliance with him and to participate in his chiefship. This marriage produced Vyogho's firstborn son, Buanga.[6]

Vyogho formed a second marriage alliance with a healer and magician named Maha, who lived in Vutungu and was an ally of Mutsora's family in Vusekuli. While this alliance may have been an attempt by Vyogho to circumvent the authority of Mutsora's descendants by forming a direct link with one of their allies, it seems unlikely that Vyogho would have risked alienating his kinsmen in Vusekuli. The alliance may therefore be seen as simply increasing the density of alliances that tied this ritually important figure to his chiefship. Maha provided Vyogho with a powerful medicine called *munhiyrwa*, which is said to have made the people of Isale love him.[7] He also served as *muhangami*, travelling with Vyogho and protecting him from pollution.

Vyogho took a third wife from the family of the rainchief Mutunzi, who lived south of the Talia at Vusali. Mutunzi's ancestors had been important rainchiefs in southern Isale before Babito settlement in the region. Their support, therefore, strengthened Vyogho's ritual control over Isale. This marriage produced Vyogho's eventual successor, Muhashu, who was born sometime just prior to the drought that struck Isale around 1897.[8]

While Vyogho depended primarily on alliances to increase his authority, there is evidence that he employed arms to extend or reinforce his control on at least two occasions early in his career. The first incident occurred soon after Kivoto's death and is said to have been initiated by the failure of the people of Kighali, who were Kivoto's subjects, to give

TABLE 7
THE ALLIES OF VYOGHO

Allies	Status	Office
Bahera of Ngukwe	senior Bahera settlement	*semombo*
Bashu of Mbulamasi	diviners of the land, rainmakers	*semombo*
Batangi of Lulinda (resettled at Ihera)	royal ritualists for early Babito chiefs	*semombo* and *baghula*
Babito of Vusekuli (Mukiritsa s/o Mutsora)	chiefs of Vusekuli; included among subjects first chiefs of Isale and Basumba of Lukanga	*mukulu*
Bito of Biabwe	healers of the land	owners of ritual seeds
Mutunzi (Vusali)	rainchief	*musingya*, presented *mbita* to Vyogho's successor
Maha (Vutungu)	healer and magician	provided protective medicines and served as *muhangami*
Descendants of Kaherataba (Mughulungu)	clearers of the forest	royal councillors (*baghani*)
Basyangwa of Muluka	clearers of the forest	*musingya*

a cup of beer to a stranger who was travelling to Kivoto's funeral. When Vyogho heard the news of this failure to give hospitality to a visitor, he detached a body of men to punish the people of Kighali. Whether or not the inhospitable actions of Kighali's residents were the real cause of this conflict is questionable, since the failure to give beer is a stereotypic explanation for conflicts in Bashu oral traditions. This is because beer drinking, despite the occasionally stormy arguments it generates, is associated with commensality, and the refusal to give beer signifies the rejection of communal ties between the person 'owning' the beer and the person who is refused beer. The story of the stranger who was refused beer may therefore indicate that the people of Kighali had in some way rejected their association with Vyogho. Vyogho's retaliation would then represent an attempt to reassert this association through force of arms. Further details concerning the exact nature of this dispute have either been forgotten or are still too sensitive to reveal to a foreign researcher. What is remembered is that the people of Kighali repulsed Vyogho's punitive expedition and thus retained their independence.[9]

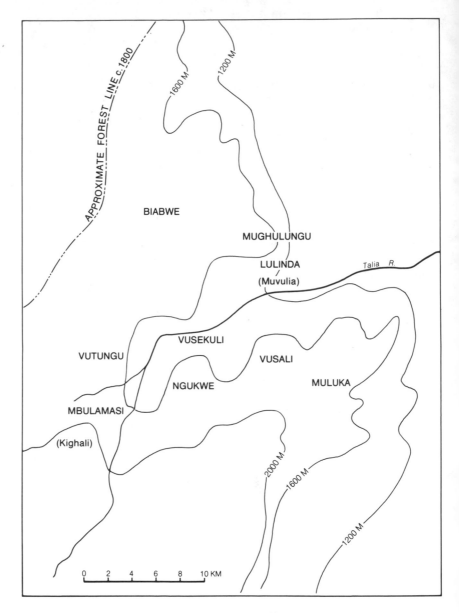

MAP 6. Vyogho's Alliances

At Muvulia, in the lower mountains just north of the Talia, Vyogho again used armed force, this time to extend his authority over an area that had formerly been independent. Muvulia, because of its location on a major line of communications between the mountains and the plains, was the site of the most important market in Isale. Here, the people of the mountains came with their products, including ivory, to trade with people coming from the plains and from Uganda. The market had apparently been independent of politics in Isale during most of the nineteenth century. However, the growing importance of trade may have encouraged Vyogho to attempt to alter this situation. Vyogho sent a client chief named Etsukura with an armed body of supporters to take control of the market.

Faced with this threat to their independence and lacking the military resources necessary to repel Etsukura, the people of Muvulia turned to the Babito chief of Mughulungu, Kivere, for military assistance. Kivere sent a military force under the leadership of a man named Muhanzangi to assist the people of Muvulia. Muhanzangi drove Etsukura away from Muvulia, and the market of Muvulia remained free of outside control until the Belgian occupation of 1923.[10]

While Vyogho's early military adventures ended in defeat, he went on, as we have seen, to establish a wide sphere of ritual influence in Isale. This success placed him in a strong position vis-à-vis both commoners and Babito kinsmen and made him a major political and ritual figure in Isale during the 1890s.

KASUMBAKALI AND THE POLITICS OF ARMED CONFRONTATION

On the surface, Kasumbakali's rise to power differed markedly from Vyogho's. Unlike Vyogho, Kasumbakali had not been designated as his father's successor. Instead, the *basingya* chose to invest Kasumbakali's younger brother, Biabo. The choice of Biabo rather than Kasumbakali is said to have been made because Kasumbakali was perceived as being too aggressive, overbearing, and ill mannered. It is also said that he insulted several members of the *basingya*. It is, of course, possible that this assessment may be based on Kasumbakali's later actions and that other factors were involved in the decision.[11] In any case, having failed to achieve power through normal channels, Kasumbakali, like Mayi among the Baswaga, turned to the alternative path to political power offered by the possession and use of firearms. To take advantage of this alternative route,

Kasumbakali formed an alliance with Tsombira, who, as noted above, was an ally of Karakwenzi's and had used firearms to increase his own authority and to conduct raids against northern Isale. Tsombira lent Kasumbakali a large number of followers, some of whom were armed with firearms. In return for this support, Kasumbakali promised Tsombira a share of the spoils that his raiding would produce.[12]

Having obtained the support of Banisanza mercenaries, Kasumbakali launched what is generally described by the Bashu as a reign of terror, pillaging villages and killing or maiming hundreds of people. Kasumbakali's raids, which occurred intermittently from 1894 to 1897, gained him wide notoriety.[13] Thus, a man from the Baswaga chiefdom of Bulengya, to the west of Isale, stated that "Among all the Baswaga, no man has killed more people than Mayi, but in the whole world, no one has killed as many people as Kasumbakali in Isale" (Bergmans, 1974: 29).

While Kasumbakali's activities contrast sharply with Vyogho's more peaceful policy of alliance formation, a closer examination of the overall pattern of his raids reveals that, despite his control of military resources, Kasumbakali continued to operate within the context of Bashu ideas concerning the political necessity of coordinating ritual activity and creating a sphere of ritual influence over Isale.

Kasumbakali's raids within Isale were by no means indiscriminate. He appears instead to have been selective in choosing his targets, using military force primarily to eliminate rival Babito chiefs and to intimidate their ritually influential allies into supporting his own chiefship. Thus, he began by attacking his brother's homestead, driving Biabo out of the area and forcing him to seek refuge in the plains. Having done so, he claimed the chiefship for himself.[14]

Kasumbakali then turned his attention to his major Babito rival, Vyogho. He attacked Vyogho's home at Vuhovi, forcing him to seek refuge among his affines, the family of Kaherataba on Mount Mughulungu. Later, by attacking the Mughulungu area, Kasumbakali forced Vyogho to depart from Isale and take refuge in Maseki, where he was hidden by the *mwami* of Maseki, Lukanda, who was the son of his paternal uncle, Kahese. This raid also caused Kamabo, the Babito chief of Mughulungu, to flee to the plains.[15]

Kasumbakali turned next to the important ritual leaders and former chiefs who had been the allies of his Babito rivals, pillaging their homes in order to force them into supporting him.[16] At Ngukwe, Kasumbakali burned the village of the Bahera chief Vyakuno, who had provided Vyogho with his *mombo*. He also took prisoner a number of women.[17]

In Vusekuli, he came looking for Vyogho's *mukulu*, Mukiritsa. When he could not find Mukiritsa, he cut off the hands of one of Mukiritsa's wives, the mother of Mukiritsa's successor, Murondoro.[18] In doing so, he demonstrated his power to the people of Vusekuli as well as to the Bito of Biabwe, who were important 'healers of the land' in northern Isale, for it was from this group that Murondoro's mother had come.[19] It is said that Kasumbakali often maimed his victims, cutting off their hands or occasionally, in the case of women, their breasts. Mutilated in this way, his victims served as grim reminders of Kasumbakali's power.

Kasumbakali killed another of Vyogho's allies, the healer Maha, who, as noted above, had given Vyogho a special medicine that made people love him. Maha's descendants claim that Kasumbakali killed Maha because "he did not want Vyogho here." Other attacks were launched against the homes of members of the *basingya* who had elected Kasumbakali's brother as *mwami* and who were also ritual leaders. These included raids in Bunyuka, Manighi, and Katikale.[20]

In contrast to the experience of these ritually important groups, the people of Muvulia, Mwenye, and Vutungu, who were also supporters of Kasumbakali's Babito rivals but who had no particular ritual or political importance, were for the most part spared Kasumbakali's attacks. The people of Mwenye in fact looked upon Kasumbakali as something of a hero, for he protected them from an attack by other groups.[21] Kasumbakali can thus be seen to have used his military force selectively, though not exclusively, to gain the support of ritually influential groups and to eliminate potential rivals for this support. His goal, then, was the same as Vyogho's, i.e., the creation of a sphere of ritual influence over Isale.

On the other hand, Kasumbakali did not limit his raids to Isale but attacked neighboring regions as well. His arena of political activity therefore exceeded the political, ecological, and ritual boundaries that appear to have shaped the previous geographical limits of political action among the Bashu.

A comparison of the traditions from areas Kasumbakali raided within Isale with those from areas he raided outside Isale, however, suggests that these boundaries did play a role in shaping Kasumbakali's political activities. In Isale, Kasumbakali's raids were, as we have seen, directed primarily against rival chiefs and ritual leaders and were accompanied by demands for tribute and recognition. His raids outside Isale were much more indiscriminate and involved attacks against all categories of people. Moreover, there is no mention of demands for tribute or recognition or of the use of maiming, a favorite method of intimidation, in the traditions of neighboring areas. The main objective of these external raids in fact seems to have been the collection of booty rather than the expansion of political authority. The distinction between these two spheres of activity suggests that Kasumbakali's actions conformed to existing ideas about the spatial limits of political authority and that his raids outside Isale were designed to support his mercenary followers and thus his political position within Isale. It is perhaps significant that this distinction between the core area of political control and peripheral raiding zones was apparently also made by Kasumbakali's patron, Tsombira, and by Tsombira's patron, Karakwenzi.[22]

THE WARS OF KASUMBAKALI AND THE FAMINE OF VYOGHO

While Kasumbakali's tactics were initially effective in establishing him as the dominant chief in Isale, they were, in the long run, unsuccessful. Most of the influential ritual leaders of the region, like their former Babito chiefs, chose to flee Isale, taking refuge in neighboring regions for fear of being killed. They thus denied Kasumbakali the ritual support he needed to control the well-being of the land and thus to establish his ritual legitimacy in the eyes of the people of Isale. This in itself would not have been critical to his position, had he not also begun to lose his control over military resources.

Kasumbakali's raids followed closely on the heels of those by the Banyoro and Manyema and soon exhausted the economic resources of Isale. This economic decline was hastened by the onset of a major drought between 1897 and 1898.[23] This drought prevented the Bashu from growing new crops to replenish their food supplies. It also killed off the few livestock that had managed to escape previous raids. The combined impact of raiding and drought on the livestock population of Isale is reflected in the statements of many people born around 1900 that they did not see a single goat in Isale until they were eleven or twelve years old. It is, in fact, commonly accepted by many of these people that the origin of goats in Isale dates from this period![24]

The raiding, drought, and subsequent famine so impoverished the Bashu region that Kasumbakali's Banisanza mercenaries no longer found it profitable to support his cause. They therefore withdrew their support and returned home to Lisasa. The loss of Banisanza support deprived Kasumbakali of his access to firearms and of his military superiority in Isale. Having lost this advantage, he was unable to maintain his control through force of arms.

As the drought continued, moreover, the people of Isale rejected Kasumbakali and began to call for Vyogho's return and to look to Vyogho as the only hope for survival. To understand this turn of events, it is necessary to examine how the Bashu themselves must have viewed the changes during this period.

From an outside perspective, the famine that hit Isale at this time was caused by forces that were independent of events occurring in Isale, i.e., changes in the wider economic and political environment of the upper Semliki Valley and a climatic anomaly resulting in an extended period of dry weather. The people of Isale, however, ascribe both the drought and the raids to a breakdown in the coordination of rituals designed to regulate the relationship between homestead and bush, and to the penetration of untamed forces of the bush into the world of the homestead.

While after Luvango's death the Babito had succeeded in maintaining their coordination of ritual activities, and thus the integrity of the sphere of ritual influence that Luvango had established in Isale, this sphere of influence was disrupted by the succession struggles that followed the deaths of Molero and Muhoyo. Both struggles resulted in the alienation and desertion of important ritual leaders who refused to support the winners of these struggles and thus prevented the victors from coordinating ritual activity within their respective chiefdoms. The Babito sphere of ritual influence in Isale was finally shattered by Kasumbakali's raids, which, as we have seen, drove many of the important ritual leaders out of Isale and, according to several witnesses, curtailed the performance of land rituals.

This breakdown in ritual control not only brought on the drought but also was seen as responsible for the raids that ravaged Bashu villages. The raiding activities of the Nyoro army, Karakwenzi, Kasumbakali, and the Manyema were a new and devastating experience for the Bashu. Whereas feuds involving the use of arms had occurred between various Bashu groups in the past, they were limited in scale and in the number of deaths that they produced. Deaths that did occur, moreover, were eventually compensated for through blood payments. In addition, feuds had traditionally evolved out of violations of social norms and therefore were explainable within the context of daily experience. Violence was thus ordered and comprehensible. In contrast to this experience, the warfare of the 1890s reached new and unprecedented levels of violence and was accompanied by all manner of atrocities that violated Bashu morality, including the maiming and mutilation of victims. Many people were killed and no compensation was paid. Moreover, the raids were beyond the logic of Bashu social experience. They were not stimulated by real or even imagined violations of social norms. They thus came without warning or explanation. They were, in short, unpredictable sources of misfortune, more like violent storms or the ravages of wild animals than the actions of men. And this is precisely how they were perceived or came to be perceived when resistance proved useless. These were not normal men but untamed forces of the bush penetrating the world of the homestead. This equation between raiding groups and forces of the bush is evident in the descriptions given by survivors of these raids and their descendants. Thus, an eyewitness to Karakwenzi's raids in Maseki described Karakwenzi's followers as "very strange men. They stole all and destroyed huts. They killed all of the men they came across and did not take prisoners. They ate the roots of gourds uncooked and drank resin oil."[25] While much of this description may describe actual events, the references to eating uncooked roots and drinking resin oil are part of a negative stereotype used elsewhere by the Bashu to describe the habits of sorcerers, *baloyi*, whose actions are antithetical to the well-being of the homestead and are identified with the forces of the bush, and *bakumbira*, men who have committed a heinous antisocial action such as incest and by doing so become creatures of the bush, living in the forest, never bathing, letting their hair grow long, eating uncooked foods and drinking resin oil.

A similar stereotype can be seen in this description of Kasumbakali's men by the son of an eyewitness: "His men killed all of the men they found here and cut off the breasts of women. They had long hair and ate only uncooked foods. They never slept in huts but in the open like animals. They slept with anyone, even their sister or mother."[26] There are, as well, the references to Tsombira as an elephant ravaging the gardens of his victims, which may be seen as a clear example of the tendency for the people of Isale to associate the groups that raided them with the world of the bush.[27]

The tendency of Bashu informants to describe invaders with images associated with the forces of the bush, whether stimulated by the actual practices of these intruders or not, suggests that the Bashu interpreted the drastic changes that occurred during this period in terms of existing ideas about homestead, bush, chiefship, and misfortune. This interpretation of misfortune helps explain why the people of Isale turned to Vyogho and rejected Kasumbakali at this time. Having explained their misfortune in terms of the breakdown in ritual control over the land, the logical solution to their plight was the reestablishment of a unified sphere of ritual control over Isale. Thus Kasumbakali, who had through his actions disrupted ritual control, was rejected and eventually driven out of Isale, to be replaced by Vyogho, who had previously established a wide network of alliances with ritual leaders in Isale and had gone a long way toward reconstructing Luvango's sphere of authority.[28]

Vyogho's credibility as a powerful *mwami*, and thus his selection as the one chief who could save the people of Isale, was further heightened during the course of the famine and following the expulsion of Kasumbakali by two developments. The first was Vyogho's introduction of the Nyavingi cult into Isale. Vyogho is said to have sent emissaries into Rwanda during the famine to acquire knowledge of Nyavingi and to bring Nyavingi back to Isale. Whether these messengers actually travelled to Rwanda, as the tradition states, or whether Rwanda simply refers to the general association of Nyavingi with Rwanda is unclear. The emissaries' descendants could only confirm that their fathers took the road south to Kyondo in Maseki and that when they returned they claimed to have been to Rwanda.[29] While this may indicate that they actually travelled to Rwanda or, more likely, to the region just south of Lake Edward around Rutshuru, which was at the time dominated by Rwanda, they may only have travelled as far as Maseki. As noted in chapter six, Karakwenzi established a large encampment in Maseki near Kyondo and may well have built shrines to Nyavingi. The inhabitants of Maseki deny this and claim instead that Vyogho was the first Bashu chief to have Nyavingi.[30] While this may be so, it is curious, given the fact that Karakwenzi spent several years in the Maseki area. On the other hand, Karakwenzi brought death and destruction to the people of the region and this experience may have prevented the cult from taking hold among them.

It is also possible that Vyogho acquired Nyavingi at Kyango, located at the northwest corner of the lake, where Karakwenzi evidently did build shrines for Nyavingi. If so, it is somewhat ironic, for it would mean that Karakwenzi, who was indirectly the source of Kasumbakali's firearms and thus of his initial success in Isale, was also the source of the ritual power that eventually helped Vyogho defeat Kasumbakali.

Whatever the source of Vyogho's Nyavingi may have been, the people of Isale had probably already heard of Nyavingi by the time Vyogho constructed his first shrine to Nyavingi *(ngoro)* at the home of his *mukulu*, Mukiritsa. Trading connections with Katwe and social contacts with the people of Maseki would undoubtedly have provided the people of Isale with news of Karakwenzi, the descendant of Nyavingi who had helped to drive the Banyoro away from Katwe and then attacked the people of

Maseki with weapons that could "kill a man in an instant" (a reference to Karakwenzi's firearms). Stanley (1890: 284), in fact, had heard of the Wanyavingi on the western sloped of the Ruwenzoris in 1889, at least eight years before Vyogho built his first Nyavingi shrine. Thus, when the people of Isale learned that Vyogho had built a hut in which he could communicate with Nyavingi and that consequently he too had the power of Nyavingi, they were understandably impressed. It is, in fact, generally believed that it was Nyavingi that enabled Vyogho to end the famine, by providing him with the power needed to restore the balance between the world of the homestead and the untamed forces of the bush. In this regard, it is interesting to note that Nyavingi was associated with both military power and fertility among the Bashu, as it evidently was in Kigezi. This dual association is consistent with the Bashu tendency to view both drought and outside invaders as untamed forces of the bush invading the world of the homestead. From an external view, of course, Nyavingi can be seen as augmenting Vyogho's ritual legitimacy in the eyes of the people of Isale and thus as strengthening his position as the one chief who could restore order and well-being.

The second development that appears to have increased Vyogho's legitimacy was the plague of jigger fleas *(Tunga penetrans)* that descended on Isale during the drought. Jigger fleas entered East Africa for the first time via caravan traffic from Zaire across Lake Tanganyika at the end of the nineteenth century. The human suffering and loss of life produced by the insect, for which no defense was known, were described by numerous European travellers to the region. Baumann, describing the effects of the flea around Uzinza in Tanzania, writes:

> Those who keep the feet clean and look after them daily to extract the jiggers have little to fear from this plague. But left to themselves, the sand-flea larvae will grow to the size of a pea and finally break out into sores. When these appear in large numbers, they can cause blood poisoning and death. Particularly in areas where the sand-flea occurs for the first time, and where its treatment is unknown, its impact can be devastating. We saw people in Uzinza whose limbs had disintegrated. Whole villages had died out on account of this vexation. (Kjekshus, 1977: 135)

The jigger flea reached the west shore of Lake Victoria by 1894 and apparently arrived in Isale via the Semliki Valley by 1897–98. The flea thrived in the dry conditions created by the drought and ravaged the population of the region. The impact of this new disaster on the Bashu is indicated by references to the famine of this period as the *nzala y'evisimi,* the famine of the jigger fleas. Vyogho is generally credited with having brought the jigger fleas to Isale in order to demonstrate his ritual power, and he may, in fact, have claimed credit for the disaster as a way of encouraging recalcitrant groups to accept his authority. In any case, the jigger fleas appear to have augmented Vyogho's legitimacy in the eyes of the Bashu.[31]

Vyogho's acquisition of Nyavingi and the onset of jigger fleas can thus be seen as having increased the strength of his claim to chiefship and contributed to his emergence as the one chief who could save the people from the disasters they faced. Consequently, groups from throughout Isale threw their support behind Vyogho's chiefship and worked to reestablish ritual control over the land under his leadership.

It could, of course, be argued that this unification was stimulated by the need to create a more effective military force to cope with the external raids that were plaguing Bashu society. Yet there is no evidence of wider military coordination resulting from this unification. At no time did Vyogho or his councillors attempt to organize a military force to cope with these raids. There is, in fact, a surprising absence of any tradition of resistance during this period. It may be that the raiding forces were too well armed and organized to make resistance possible. In addition, the Bashu may have been so decimated from repeated raids that by the time unification was achieved they simply lacked the physical strength to organize a concerted resistance. There may, however, be an additional reason for this lack of military response. Given the nature of the raids and the way in which they came to be perceived, i.e., as untamed forces of the bush, resistance through military means may have seemed inappropriate. One could no more use armed force to combat the Manyema or Karakwenzi than to fight malevolent spirits of the bush or to deter a thunderstorm.

In the end, the drought gave way to wetter conditions, which reduced the problem of jigger fleas. Moreover, Kasumbakali's loss of armed supporters, and British and Belgian activities in the upper Semliki Valley, reduced the level of banditry and violence in the mountains. With these changes, peace and relative prosperity slowly returned to the Bashu region, allowing people of the region to rebuild their homes and lives. From an external view, this change in circumstances occurred independently of actions taken by the people of Isale. Nonetheless, the Bashu saw the end of famine as a logical consequence of the reestablishment of ritual control and the creation of a widespread sphere of ritual influence over the whole of Isale, just as the onset of famine had resulted from the disruption of ritual control. External events, therefore, verified Bashu explanations for the disasters and in doing so reinforced the conceptual framework upon which these explanations were based. Consequently, the traumatic events of the last years of the nineteenth century can be seen as having reaffirmed rather than undermined Bashu political ideas.

FAMINE AND THE EXPANSION OF BASHU CHIEFSHIP

Yet it would be a mistake to conclude that the events of this period had no effect on Bashu political perceptions, for the period saw a redefinition of the geographical limits of political authority. Before this period, the limits of chiefship appear to have been defined by a combination of ritual, ecological, and political factors that inhibited the expansion of political authority beyond the borders of Isale. The raiding, drought, and jigger fleas of the 1890s, however, affected the whole Bashu region, reducing the significance of ecological and political divisions and encouraging the expansion of ritual cheifship.

Faced with the desperate conditions of the period, people looked initially to their existing chiefs for relief. As the situation continued to deteriorate, however, many Bashu began to lose faith in their local leaders and to look further afield for any chief who might be able to reestablish ritual control and bring an end to their troubles. In these circumstances, Vyogho's previous success in creating a widespread sphere of influence in Isale, his possession of Nyavingi, and, perhaps most important, his acquired status as the rightful successor to the earliest Babito chiefs in the Bashu region—from whom most of the other Babito chiefs among the Bashu were descended—soon gained him the support of chiefs and commoners from throughout the Bashu region. The only major exception to this pattern was the Babito chief of Malio, Mbonzo, who allied himself with Kasumbakali and refused to acknowledge Vyogho's authority.

The support of these outlying groups was not in itself sufficient to end the famine, however. It was also necessary to convert this support into a unified sphere of ritual influence. To accomplish this, the temporal dimension of Vyogho's chiefship had to be extended over chiefs and ritual leaders throughout the region. This was achieved in several ways. First, Vyogho underwent a second investiture ceremony, in which he went into seclusion and was anointed again with red powder and resin oil and put on the *mbita*. This process renewed and strengthened the

chiefship and resembled the 'new moon' ceremonies that are held periodically for this purpose. Vyogho's reinvestiture differed from the 'new
moon' ceremonies, however, in that it was preceded by the investiture,
or reinvestiture, of all the chiefs and ritual leaders of the expanded region
of his ritual control. This procedure brought the developmental cycles
of these ritually influential offices into line with that of *bwami* and insured
that their ritual power was renewed and brought to full strength along
with Vyogho's power.[32]

Second, planting rituals marking the beginning of the agricultural year
were coordinated to occur at the same time throughout the region. This
coordination was difficult to achieve because of differences in the timing
of agricultural cycles within this wider region. As noted previously, these
rituals traditionally began before the planting of eleusine, which the
Bashu say was the first, in the sense of being the oldest, crop. Yet, because
eleusine was planted in September in Isale and in February in the higher
regions to the south, the central new year planting rituals occurred at
different times in the two areas. To overcome this obstacle, the timing of
the new year rituals in the southern areas was altered to conform with
their performance in the north, and specifically with the ritual observances directed by Vyogho in Isale. Thus, instead of occurring before the
planting of eleusine, they now occurred before the planting of beans, and
thus in September, in conjunction with ceremonies in Isale. Existing ideas
about the natural order of agricultural events were thus subverted to
allow for the creation of a wider sphere of ritual influence. This alteration
can be seen as a further example of the way in which the temporal
dimension of Vyogho's chiefship was extended over outlying areas.[33]

The process of ritual unification and coordination was further extended by the geographical expansion of the annual cleansing ritual that
accompanied the new year ceremonies. Prior to this period, cleansing
rites had been performed annually by each village, with the villages on
a single ridge coordinating their activities. Under Vyogho's direction,
these local cleansings were joined together in a unified cleansing of the
whole area of Vyogho's control. A special group of *bakumu* travelled
throughout the region collecting the bunches of herbs that had been used
to cleanse each ridge. They began in northern Isale and ended at the river
Tumbwe, which presently marks the southern limits of the Bashu region.
Here all the ritual brushes were thrown in the river, and offerings were
made to keep the malevolent spirits and other uncontrolled forces of the
bush away from the homestead, defined in this context as the entire area
of Vyogho's control.[34]

As a result of these innovations, Vyogho was able effectively to extend his authority beyond the limits of Isale, thereby expanding the ritual and political scale of Bashu chiefship. Over the next twenty years, from approximately 1900 to Vyogho's death in 1922, Vyogho gradually increased his authority over this wider region. This process was facilitated by several factors. First, and perhaps most important, the events of the preceding decade had confirmed the value of maintaining centralized ritual control over the region. Thus Bashu experience reinforced the ideological foundation for centralized chiefship. Second, the first two decades of the twentieth century were years of relative peace and prosperity for the Bashu. In contrast to the southern Mitumbas and the western slopes of the Ruwenzoris, where the Belgians launched major pacification campaigns prior to World War I, the Bashu had only limited contact with Belgian forces (Joset, 1939: 13–16, 60–68). Belgian activities were limited to the sporadic raids conducted by their African ally Kirongotsi, and to occasional visits by reconnaissance parties seeking the assistance of local chiefs in acquiring food supplies and porters for their posts in the Semliki Valley.

Natural conditions, while somewhat less conducive to prosperity, did not drastically disrupt Bashu society. With the exception of a period of dry weather in 1911 that affected northern Isale (Joset, 1939: 14), there is little evidence of climatic crises during this period. The outbreak of human sleeping sickness in the Semliki Valley during the first decade of the century was a more serious problem. The disease ravaged the populations that lived along the tsetse-infested riverine areas that transected the valley. Moreover, Belgian resettlement schemes, which had created consolidated villages near these rivers, greatly increased the number of deaths that resulted from the disease (Ford, 1971: 177–78).

While the disease itself did not affect the health of peoples living in the higher altitudes, it did affect the Bashu economy, which, as we have seen, was tied to that of the plains through trading relations. The disease increased the dangers inherent in travelling to the plains and thus reduced the salt trade. Those who continued to travel to Katwe first had to visit a *mukumu* to obtain, and pay for, protective medicines against the disease.[35] The disease also forced the pastoral groups, on whom the Bashu depended for livestock, to move away from the region closest to the mountains, which were heavily infested with tsetse, and to settle to the east of the Semliki. This made access to livestock more difficult at a time when the livestock population of the mountains had been decimated by raids and drought. Consequently, the goat population in the mountains did not begin to increase until the beginning of the second decade. The disease did not, however, wipe out the livestock population of the valley, as it did in other regions of East Africa. Throughout the early years of the century Belgian reports record the presence of significant herds of cattle and other livestock among the peoples of the plains. Had these animals been killed off, the disease would have had much more serious consequences for the livestock-starved peoples of the mountains.

The real impact of the disease was not felt until after Vyogho's death and the Belgian occupation in 1923. It was, in fact, Belgian attempts to control the disease through the systematic evacuation of the valley and lower mountain slopes, rather than the disease itself, that had the greatest impact on the Bashu. For this policy cut the Bashu off from important agricultural areas located in the lower mountains and from traditional markets in the plains.

In general, then, the first two decades of the twentieth century were years of relative peace and prosperity. These conditions, when viewed against the backdrop of the disastrous events of the previous decade, justified and encouraged the continued centralization of ritual control and thus reinforced Vyogho's authority. Moreover, the return of plentiful harvests, combined with the expanded geographical scale of his political authority, brought Vyogho large quantities of food supplies as tribute. This surplus was used to reinforce his position through redistribution to his subjects. Thus, Vyogho is remembered as a generous chief, "who never ate alone."[36] The food supplies were also used for exchange along with ivory, which continued to be an important trade item during the early decades of the twentieth century. Vyogho, who as *mwami* possessed a partial monopoly over ivory collected within his chiefdom, was a major participant in this trade. One informant remembered having visited Vyogho's home after the famine and having seen many tusks in a hut in Vyogho's compound. Through trade Vyogho acquired a variety of goods, including imported cloth, beads, and hoes. Many of these items were redistributed to loyal supporters. Hoes were a particularly valuable gift during the first years of the century, because, in the absence of goats, hoes became the primary basis for Bashu bridewealth. Vyogho later acquired goats and some cattle through his trading relations. The scarcity of these animals made them an important political resource. Vyogho's ability to acquire livestock and to provide his supporters with them, combined with the virtual elimination of livestock during the previous decades, resulted in the widespread belief that it was Vyogho who first introduced goats to the mountains.[37]

With both a strong economic base and ideological supports, Vyogho was able to increase both his ritual and political authority in Isale and, to a lesser degree, in neighboring regions. Ritually, Vyogho succeeded in forcing certain 'healers of the land' within Isale to relinquish control of the ritual seeds, *mikene*, which they had previously distributed at the annual planting ceremonies. Starting with Vyogho, these healers supplied the seeds, but Vyogho distributed them to his subjects. In this way, he maintained the ritual influence of the *bakumu* but increased his direct control over the use of this ritual resource.[38] It is currently maintained that the *mwami's* temporary possession of the *mikene* heightens their power and is essential to their successful use. Similarly, while Vyogho was unable to monoplize the use of rainmagic, he is said to have been able to prevent rainmakers from using their rainstones without his permission, though how this was done is unclear.

Vyogho further strengthened his personal control of ritual resources by institutionalizing the veneration of Nyavingi, the spirit that had helped Vyogho end the famine of the jigger fleas. Under Vyogho's direction Nyavingi came to play a central role in all ritual observances for the land, and shrines for Nyavingi *(syongoro,* sing. *ngoro)* were constructed at the homes of each of Vyogho's subchiefs. To complete the incorporation of Nyavingi into Bashu royal culture, Nyavingi was transformed from a borrowed alien spirit into a female ancestor of the Babito.

Politically, Vyogho succeeded in establishing a hierarchy of subchiefs in Isale, composed of representatives of collateral Babito lines whose direct ritual and political authority over particular areas of Isale he recognized, and of members of his own lineage, whom he placed over the remaining areas of Isale. These chiefs *(bakama)* initially served as ritual intermediaries between Vyogho and the people of Isale, conveying the people's needs to him and his blessings to the people.[39] They were also responsible for organizing local ritual activities and coordinating them with those performed by Vyogho. With time, these chiefs took on certain political and judicial functions within the communities over which they ruled, deciding cases and collecting tribute, which was paid to Vyogho. The distribution of these chiefs, however, was uneven, and certain areas, occupied by particularly important allies, remained outside this chiefly hierarchy. (The Bashu of Manighi and the Baswaga of Katikale, for example, were never required to accept such a subchief.) Vyogho himself began to acquire appellate jurisdiction over Isale, deciding cases that could not be resolved at the local level.

Beyond Isale, in Maseki, Kakuse, Kasongwere, and Ngitse, Vyogho exercised only indirect authority. The chiefs in these regions continued to maintain direct political and ritual control over the land, although in the realm of ritual they were, as we have seen, required to coordinate their activities with those of Vyogho. There is no evidence that Vyogho was able to force ritual leaders in these regions to relinquish their ritual resources, nor, apparently, were local chiefs able to do so on their own. There is similarly no indication that Vyogho extended his hierarcy of client chiefs over these outlying areas. Vyogho did, however, receive a portion of the tribute paid to local chiefs on a regular basis.

Finally, Vyogho's reputation as a powerful chief reached beyond the limits of the Bashu region to other Nande groups to the south and west, whose chiefs regularly sought his assistance in ritual matters, and as far east as Katwe, where Vyogho's ritual assistance was sought on at least one occasion, when adverse climatic conditions curtailed the production of salt in the lake. Vyogho's relationship with these areas, however, remained informal, and no attempt was made to extend the temporal dimension of his chiefship over them or to coordinate the performance of ritual within this wider area.[40]

In summary, Vyogho can be seen to have established three concentric circles of authority. The core of his authority lay in Isale, where his political and ritual powers were extensive, allowing him to achieve a greater degree of ritual consolidation than had been accomplished by any of his predecessors and permitting the establishment of a hierarchy of chiefs. Isale was, in fact, beginning to resemble some of the kingdoms of the interlacustrine region, albeit on a much reduced scale. Beyond Isale, in the southern reaches of the Bashu region, Vyogho possessed considerable ritual authority and had extended the temporal dimension of his chiefship over local ritual leaders and chiefs, subordinating the developmental cycles of their offices to that of his own chiefship. Local chiefs, however, still performed their own annual planting rituals and were directly responsible for coordinating ritual activities within their chiefdoms. They were also responsible for the placement of their own client chiefs. In more distant regions, Vyogho's status was like that of a distant oracle whose advice and assistance was periodically sought but who had no regularized local authority.

CONCLUSION

In comparison to the small-scale rainchiefdoms that had existed among the Bashu when the Babito first settled in Isale at beginning of the nineteenth century, Bashu society on the eve of colonial rule had achieved a remarkable degree of centralization and unity. Moreover, chiefs had taken on more secular responsibilities. Yet the basis of chiefly authority remained firmly based on the possession of ritual legitimacy. The events of the previous thirty years had confirmed Bashu political ideas, and the expanded political system that emerged under Vyogho's leadership, while facilitated by new economic resources provided by the ivory trade, resulted, in large measure, from the application of Bashu ideas about chiefship and the need for centralized ritual control to the changed economic and political circumstances of the 1890s. For it was the interpretation of these circumstances in terms of Bashu ideas concerning the homestead, bush, and chiefship that legitimized and encouraged the expansion of Vyogho's chiefship.

The advent of formal colonial rule in the 1920s presented an even more serious challenge to the ideology of ritual chiefship. For, as elsewhere in Africa, colonial rule created new sources of legitimacy and authority that were largely independent of traditional sources of political authority and challenged their validity. While these changes eventually altered the nature of Bashu chiefship as well as patterns of competition, Bashu political ideas continued to exhibit a remarkable persistence in the face of change, and by doing so continued to color political developments during the colonial period.

CHAPTER EIGHT

COLONIAL RULE AND THE POLITICS
OF RITUAL CHIEFSHIP

THE ESTABLISHMENT OF BELGIAN RULE

By 1900 the Belgians had established their political authority over the upper Semliki Valley, founding administrative posts at Beni, Kasindi, and Katwe and eliminating the opposition of the Manyema and Karakwenzi. From their posts in the valley, the Belgians expanded their grid of administration over the surrounding mountains during the first two decades of the twentieth century.

In the Bashu region, the process of pacification began with a series of reconnaissance expeditions designed to make contact with the mountain chiefs and gain their assistance in procuring agricultural produce and porters for the Belgian posts in the Semliki. These efforts met with mixed success. While some chiefs responded favorably, many refused to have any relations with the Belgians.[1]

The outbreak of sleeping sickness in 1905 and the coming of the First World War reduced Belgian manpower in the region and prevented more active attempts to establish direct control over the Bashu until the 1920s. With the end of the war, the Belgians made a more concerted effort to gain the submission and cooperation of the Bashu chiefs and their Baswaga neighbors. In May of 1922, the Belgian territorial administrator Ransbotijn led a military expedition from the Semliki into the southern Bashu region.[2] The expedition was attacked at Maghigi, and in the ensuing battle a number of Bashu men were killed. A second battle occurred near Kyondo.[3] These skirmishes were apparently the only armed confrontations between Belgian and Bashu forces prior to the submission of the Bashu chiefs in 1924. The absence of a more extensive campaign of resistance is somewhat curious, given both the occurrence of such a movement among other Nande groups and the degree of political centralization achieved by Vyogho during the previous decade.[4] The lack of more active resistance, however, can perhaps be explained by Vyogho's death just before the Ransbotijn expedition and the political disruption that it caused.

Vyogho's death was a traumatic event that disrupted the entire Bashu region. The Bashu were, in fact, still mourning his death, and all normal activities had ceased when Ransbotijn arrived on the scene. Vyogho's death, moreover, threw into question the continued unity of the political system he had pieced together. While Vyogho had evidently designated his son Kitawiti as his successor, it was not at all certain that the various chiefs and ritual leaders who had accepted Vyogho's authority would accept that of his successor. Bashu experience with centralized political control had, after all, been relatively brief, and Kitawiti's continuance of Vyogho's chiefship was no doubt looked upon with some misgivings by other Babito chiefs. For these reasons, Bashu ability to organize a concerted resistance to Belgian incursions may have been limited.

The Belgians, for their part, did not initially take advantage of this internal weakness, choosing instead to concentrate their activities on securing a safe route of communications between their post at Beni and a second post, which had been established at Lubero in southern Bunande by Belgian forces penetrating the Nande region from south of Lake Edward. This line of communications passed to the west of the Bashu chiefdoms (Bergmans, 1974: 49–50).

By 1924, however, the Belgians had secured this north-south route and begun expanding their activities out from it. This resulted in renewed attempts to gain the submission of Bashu chiefs. In the face of this increased pressure, Kitawiti moved boldly to establish an alliance with the Belgians, agreeing to meet with the Belgian Lt. Patefoort in March of 1924. By doing so he became the first Bashu chief to submit to the Belgians and was able to press his claim to being Vyogho's rightful successor and the paramount chief of the Bashu.[5] The Belgians, who had observed the influence that Vyogho had exercised during their earlier reconnaissance expeditions, accepted Kitawiti's claim and recognized him as the *grand chef* of the Bashu region. Kitawiti thereby solidified his political position over the Bashu.[6]

RITUAL CHIEFSHIP VS. THE OFFICE OF *GRAND CHEF*

The Belgians viewed the office of *grand chef* as having a traditional basis in Bashu society, and, within the limited historical context of the first two decades of the twentieth century, there was a certain justification for this view. Vyogho had established himself as something like the paramount chief the Belgians envisaged when they created the office of *grand chef.* Yet even within this context there was a wide gap between Belgian ideas and expectations about the nature of chiefship and Bashu political perspectives.

The Belgians saw the *grand chef* as a local administrator who would assist them in implementing social and economic policies, oversee the collection of taxes, assist in the procurement of agricultural produce, and help in the recruitment of labor. All these roles, with the possible exception of tax collection, were largely alien to the Bashu conception of a chief's role. To the Bashu, a chief was first and foremost a ritual figure, and his executive powers were limited.

The failure of early Belgian administrators to perceive the ritual character of Bashu chiefship may have been based in part on prior assumptions about the political nature of chiefship. They were evidently also misled by the prohibitions that prevented a *mwami* from actually carrying out sacrifices for the land and by the role of the *mukulu* as his ritual surrogate. The Belgians concluded from this practice that the *mukulu* was the ritual chief among the Bashu and that the *mwami* was a chief executive, responsible for secular matters.[7] In the end, of course, local administrators had very little discretion in the matter, since they were under administrative directions requiring them to establish local paramount chiefs as the central focus for native administration.[8]

Whatever the reason for this misconception about the nature of Bashu chiefship may have been, once it was formed it became the basis for subsequent expectations concerning the role of the *grand chef* and for evaluating the performance of individual chiefs. In short, the *grand chef* was to be a secular authority with traditional legitimacy who would serve Belgian administrative needs in return for an annual salary. This expectation conflicted with Bashu political values and created a prolonged struggle between Belgian administrators and Bashu leaders, in which the Belgians attempted to impose their political values and perspectives onto the Bashu political system, while the Bashu attempted to maintain the original character of that system. This struggle between the advocates of conflicting ideologies—of which neither side was completely conscious —forms the central theme of Bashu political history during the colonial period and attests to the continued vitality of Bashu political values during this period.

While the conflict between Belgian administrators and Bashu notables over the definition of chiefship was inherent in Kitawiti's appointment as *grand chef,* it did not surface until after his death. There are two reasons for this. First, Kitawiti was widely regarded as having inherited Vyogho's ritual power over the land and thus his appointment as *grand chef* did not conflict greatly with Bashu political values.[9] Second, the actual disparity between his ritual functions and his role as *grand chef* did not have sufficient time to emerge, for he died in May 1925, only six months after his appointment as *grand chef.*[10]

Following Kitawiti's untimely death, the Belgians chose his eldest son, Mukanero, as *grand chef.* This action created the first open conflict between Bashu and Belgian political perceptions. From the Belgian viewpoint, Mukanero, as Kitawiti's son, was the logical successor to the office of *grand chef.* The Bashu, however, did not share this view, for Kitawiti had never undergone the traditional rites of accession. Thus, despite his inherited ritual power, he had not been confirmed as a *mwami.* His sons therefore had no claim to the chiefship, which should have gone to one of Kitawiti's brothers, Muhashu or Buanga. Consequently, at the same time that the Belgians were presenting the medal of *grand chef* to Mukanero, Bashu notables were making preparations for investing one of Kitawiti's brothers as *mwami.* In short, the Bashu chose to ignore Mukanero and, with him, the office of *grand chef.*

Mukanero was thus a chief without a constituency and had little capacity to fulfill Belgian expectations. He may have realized the futility of trying to do so, for he kept a low profile during his reign.[11] As a result of Bashu passive resistance to the office, Belgian attempts to establish a system of indirect rule around the *grand chef* met with an initial setback. The first round in the political struggle had gone to the Bashu.

The struggle was far from over, however, and if Mukanero chose not to champion the Belgian cause, there were others who would, thereby creating a serious challenge to Bashu political values. The first of these was Mukanero's younger brother, Musayi. Musayi appears from his actions to have been far more aggressive than his elder brother. Moreover, he recognized that Belgian support could be used to wrest the chiefship from his uncles. To this end, he accused Muhashu and Buanga of having poisoned Kitawiti and of obstructing the activities of Mukanero. Through these attacks, he worked to have his uncles exiled and thus to eliminate the threat posed by the imminent investiture of one of them. From a Belgian viewpoint, Musayi's allegations provided an explanation for Mukanero's failure as a chief. More importantly, it was an explanation that did not challenge their initial assumption concerning Mukanero's legitimacy. It was perhaps for this reason that they chose to accept

Musayi's allegations and subsequently exiled Muhashu and Buanga to Muhagi.[12]

The exile of Muhashu and Buanga preempted any effort on the part of other Bashu elders to invest a successor to Vyogho. It did not, however, make Mukanero any more legitimate in Bashu eyes, and he continued to be an ineffective local administrator. The Belgians were therefore forced to divest him of his position as *grand chef* in 1929.[13] As a replacement they naturally turned to his energetic brother, Musayi.

From a Belgian viewpoint Musayi's appointment brought an initial improvement in local administration among the Bashu. Musayi showed himself to be much more active and efficient in collecting taxes and in enforcing various regulations than Mukanero had been. His success, however, was due not to any increase in legitimacy but rather to his willingness to employ the African police force of twenty-five men that the Belgians had allotted him to harass the peoples of Isale and neighboring regions who refused to accept his authority. From subsequent inquiries and the testimonies of Bashu elders, it appears that Musayi employed a wide variety of methods to punish recalcitrant subjects. He stole their livestock, exacted excessive taxes, and burned their huts and, in some cases, whole villages. He also imprisoned a number of elders for periods of fifteen days, as was allowed under Belgian administrative guidelines.[14]

Musayi's activities forced Bashu notables to acknowledge the power inherent in the office of *grand chef* and to recognize that real political power no longer rested solely on ritual legitimacy. They could no longer ignore the office but would have to find a means of dealing with it. It was these realizations that led the leaders of a number of southern areas to begin a passive resistance campaign, in which they refused to pay taxes, sell crops to European buyers, or present themselves to any government official or buyer.[15] By taking this course of action they eventually forced the Belgian authorities to recognize their grievances against Musayi. Although the Belgians initially responded by ordering a second military occupation of the affected regions, they subsequently conducted inquiries into the causes of the disturbances. These inquiries revealed how intense Bashu resentment against Musayi had grown. They also indicated that a similar occupation would be required at a later date if Musayi were not removed. This created an administrative crisis. For now that the appointment of Musayi had been justified on the basis of the supposed "traditional legitimacy" of Kitawiti's descendants, there was no one else among his brothers and sons who was able to take his place. His own sons were too young, and his brothers showed no promise as effective leaders. The Belgians, therefore, had to choose between staying with Musayi and risking further unrest; appointing another member of his family and being saddled with an ineffective leader; or looking for a chief in a collateral line, which would mean abandoning the requirement of traditional legitimacy.

Musayi's untimely death in November 1932 simplified the decision. However, the Belgians still had to choose between legitimacy and competence. The Belgians chose competence and appointed Muhayirwa, a descendant of Visalu, as *grand chef*. Muhayirwa was seen as a representative of the younger generation and as someone who had demonstrated a willingness to work with the Belgians and serve as an effective administrator. The problem that remained was to find a way to justify this choice in terms of traditional notions of legitimacy—or rather in terms of Belgian perceptions of traditional legitimacy (Joset, 1939: 113).

The solution to this problem is reflected in the official Bashu histories that can be found in administrative reports beginning in 1932. These histories claim that Visalu was the real *grand chef* and that he was overthrown by Luvango. Thus, Luvango's descendants up to and including Musayi were illegitimate usurpers, and legitimate authority rested with Visalu's line. A variant of this revision states that Muhayirwa should have been invested as chief but because he was too young the chiefship had gone to Kitawiti and then was mistakenly frozen in that family by the Belgians. While it is possible that the Belgians fabricated these traditions in order to justify their appointment of Muhayirwa, Belgian correspondence from the period just before Musayi's death suggests that the Bashu themselves had begun to disseminate these variants as a way of undermining Musayi's authority.[16] This is an interesting possibility, for it suggests that the Bashu themselves were expressing their opposition to Musayi in an idiom that conformed to Belgian misconceptions about the office of *grand chef*. After all, there had never been a paramount chief on the geographical scale of the *grand chef* prior to Vyogho's reign. In a very real sense, Vyogho had, as far as the Bashu were concerned, created the chiefship. Thus, any debate over the legitimacy of Vyogho's ancestors was irrelevant to the realities of political authority and could only have been made to discredit Musayi in Belgian eyes. As subsequent events would reveal, legitimate political authority for the Bashu continued to reside with Vyogho's successors.

Yet, if Bashu leaders did indeed disseminate traditions designed to discredit Musayi, why did they designate Muhayirwa as the 'true' *grand chef*? Why not indicate Vyogho's sons? There may have been several reasons for choosing Muhayirwa. To begin with, Vyogho's sons Muha-shu and Buanga were in exile. Thus, the Bashu may have designated Muhayirwa as a temporary replacement who would serve as *grand chef* until such time as the return of Vyogho's sons and the choice of his successor could be orchestrated. Still, the choice is curious, for Mu-hayirwa was hardly the type of chief one would have expected the Bashu to choose. He made no pretence about being a ritual leader and refused to be invested in the customary manner. Moreover, as a Catholic convert, he refused to take more than one wife, thereby rejecting traditional patterns of support-building, patterns linked to a *mwami's* role as the coordinator of ritual authority. On the other hand, this nontraditional behavior may have been precisely why he was chosen. As a secular authority lacking traditional qualities associated with chiefship, he posed no threat to the orderly succession of Vyogho's sons once their return had been achieved. The choice of Muhayirwa, a nontraditional leader, may also have been made in order to separate the office of *grand chef* from that of ritual chief and from the politics of ritual chiefship. In this context, Kitawiti's untimely death may have suggested that combining the two offices was too dangerous, opening the incumbent to pollution and death and thereby threatening the well-being of the land. Finally, it is possible that some branches of the Babito favored Muhayirwa because they wished to weaken the dominance of Vyogho's descendants. Split-ting secular and ritual authority between two different branches would certainly limit the political power of any one branch, without necessarily disrupting ritual control.

All of this assumes, of course, that Belgian reports concerning the role of Bashu elders in presenting revised traditions were not simply fabri-cated by local Belgian administrators in order to justify the choice of Muhayirwa as *grand chef.* There is no way to know for certain whether or not this was the case. Bashu informants were extremely vague about their role in Muhayirwa's original appointment, in large part, I suspect, because his son Machozi held the office of *grand chef* at the time of my research.

Whatever the reason behind Muhayirwa's appointment, it represented a temporary victory for the Belgians in their struggle to establish a secular administration over the Bashu through the office of *grand chef* and to find a candidate for the office who at least had the appearance of possessing traditional legitimacy, even if they themselves had fabricated this appear-ance. The victory was short-lived, however, for within two years, the Bashu were calling for Muhayirwa's removal and his replacement by one of Vyogho's sons.

COLONIAL DEVELOPMENT, FAMINE, AND THE REAFFIRMATION OF RITUAL CHIEFSHIP

The Belgians saw the call for Muhayirwa's removal as the product of a conspiracy formulated by a handful of Bashu subchiefs and notables who simply refused to accept any central political authority and who consequently plotted to undermine the authority of any chief the Belgians appointed. The problem was perceived in terms of a conflict between centralism and federalism.[17]

Bashu informants, on the other hand, provide a different view of Muhayirwa's downfall. From their vantage, the issue was not centralism, although some chiefs no doubt resented the political authority inherent in the office of *grand chef*, but the declining condition of the land and of society, which could only be restored through the reestablishment of ritual control and the investiture of a legitimate *mwami*.

Muhayirwa's tenure in office coincided with a period of increased social and economic upheaval created by Belgian efforts to transform Bashu society. While I do not intend to present a complete analysis of colonial development among the Bashu, it is necessary to describe some of the major changes that were brought on by Belgian rule and to examine their impact on Bashu society in order to understand why the Bashu called for Muhayirwa's removal.

With the establishment of colonial rule in the early 1920s, Belgian authorities attempted to draw the Bashu region into the periphery of colonial economic development. Corvée labor was recruited to work on local administrative projects, such as constructing roads and buildings, and to serve as porters. Other labor was recruited by the Minière des Grands Lacs to work in mines located to the north and west of the Bashu region, although mine recruitment did not create a serious demand for Bashu labor until the 1930s and 1940s.[18] The Belgians also sought to extract surplus agricultural produce from the Bashu region to support mining activities. These policies contributed to the impoverishment of the Bashu region.

The extraction of agricultural produce began even before Belgian control was established in the mountains. Thus, as we have seen, Belgian reconnaissance parties had as one of their primary objectives the acquisition of food supplies for their administrative posts in the Semliki Valley. These attempts met with mixed success, and voluntary compliance with Belgian demands was replaced by coercion after the occupation. The

Belgians, like other colonial rulers, imposed taxes payable only in colonial currency in order to force the Bashu to sell their surplus production to European buyers. Incentives were also paid to local chiefs to gain their assistance in "encouraging" farmers to supply the buyers with food.[19]

The Bashu had traditionally grown significant agricultural surpluses, a part of which had been used to pay tribute to chiefs and other local ritual leaders. A second part, in the form of sorghum and eleusine, had been stored as a hedge against famine. The remainder was used for trading with peoples from the plains and forest and for acquiring salt from Katwe.

In normal years, therefore, the Bashu had little trouble supplying European buyers with food supplies and were able to pay their taxes. The taxes, however, were an additional expense, since they supplemented rather than replaced traditional forms of taxation. To meet the new taxes the Bashu had either to allocate surpluses destined for famine reserve, decrease their trading resources, or increase their production. In most cases, it was the trading resources that were tapped, for reducing famine reserves was too dangerous, and the ability to increase production was limited by several factors.

To begin with, demands for corvée labor reduced the manpower available for clearing new land and thus for increasing food production. Production was further restricted by the evacuation of the Semliki Valley below 1000 meters as a precaution against the spread of sleeping sickness in 1932. This cut off access to gardens that peoples in the higher regions maintained below 1000 meters. Moreover, it forced people who lived below this level to move higher up in the mountains and thus increased pressure on land resources above 1000 meters.[20] In addition, 1932 saw the forced movement of the Bashu into consolidated villages for purposes of administrative efficiency and, supposedly, to improve sanitary conditions.[21] The movement into consolidated villages made it more difficult for the Bashu to protect their gardens from the ravages of wild animals, and especially the foraging activities of monkeys and baboons. Elephants also seem to have been a problem. The destruction of crops by elephants caused Bashu chiefs to request that village headmen be issued guns to protect their peoples' gardens in 1935.[22] Thus, the Bashu found it difficult not only to increase their production but even to maintain precolonial levels of production. In the end, therefore, tax obligations were largely met with resources that had previously been targeted for exchange, thereby reducing their purchasing power.

The Bashu ability to acquire trade items was further reduced by the loss of traditional markets. The evacuation of the Semliki Valley, together with the outbreak of cattle disease in 1931, forced most of the herdsmen of the plains to migrate to Uganda, cutting the Bashu link to important trading partners, and specifically to their primary source of livestock. The evacuation of the valley also curtailed the trading activities of Ugandan merchants and greatly restricted Bashu access to salt, which was not only consumed locally but was also used for trading purposes within the mountains.[23] Moreover, salt that was available was more expensive, both because of the difficulties in acquiring it and because of the British imposition of a tax on salt production. In short, the Bashu lost many of the traditional markets at the same time that available trade resources were declining. While new marketing structures were eventually established in the mountains in the mid-1930s, there were until this time few opportunities for acquiring the imported goods that the Bashu had formerly gotten through trade.

As if these problems were not enough, Bashu productivity and buying power were further reduced by two additional events that occurred during Muhayirwa's rule. The first was the worldwide depression, which hit the Bashu region in the early years of the 1930s. This resulted in declining mine production and a reduction in the demand for agricultural produce to feed mine labor. This decline in demand initially led to a decrease in the price paid for Bashu produce. This meant that a larger share of their surplus production and ultimately a portion of their subsistence production had to go for paying taxes, which remained at predepression levels. By the end of 1932, the mines ceased buying Bashu produce altogether, leaving the Bashu unable to pay their taxes.[24] The second blow to hit the Bashu region was a plague of locusts, which destroyed a large portion of the crops in Isale in 1935.[25]

As a consequence of these various factors, the material standard of living among the Bashu appears to have suffered a substantial decline during the first decades of colonial rule. It is, of course, difficult to measure this decline in absolute terms, for I have no data on precise levels of production and trade during the precolonial period and only limited data for the early colonial period. Yet, if one listens to what the Bashu themselves say about their economic situation during the early years of colonial rule, it is clear that the period was perceived as one of steady economic decline when compared with the years of relative prosperity just prior to the Belgian occupation. In simple terms, where a farmer might have been able to convert his agricultural surplus into several new hoe blades, a goat or two, and an assortment of other imported and locally produced goods during a good year prior to the establishment of Belgian rule, very few of these items could be acquired by the early 1930s, and there were shortages of food. One informant, describing the early colonial period, summed up the problems facing Bashu cultivators during this period in the following statement: "We had to use hoe blades which were mended several times and when these were broken, we used wooden sticks to clear the land."[26] Thus, declining resources and markets reduced access to the means of production, which further curtailed cultivation in an ever downward spiral.

As bad as economic conditions appear to have been, they were not the only problems facing the Bashu during the early years of Belgian rule. The Bashu also claim that the movement into consolidated villages caused an increase in sickness and deaths. While it is difficult to document this claim accurately with existing demographic and medical records, there is evidence that two diseases in particular may have been more prevalent within the consolidated villages: dysentery and plague. Dysentery was evidently a problem from the beginning, as there are several references to antidysentery measures being taken by Belgian administrators during the early 1930s.[27] The first reference to plague that I could find occurs in an A.I.M.O. report for 1941. The report states that "The Bashu chiefdoms have once more had to mourn several deaths by disease (plague)."[28] The statement implies that earlier cases had occurred, although how early these were cannot at present be determined. Inadequate sanitation in conjunction with an increase in the density of settlement could well have contributed to a sharp increase in the incidence of both diseases.

The consolidated villages also caused psychological hardships. People had to learn to adjust their lifestyles and to coexist with one another within the close confines of the village world. Tensions and conflicts arose over the placement of huts, the use of limited resources of land and water, and the loss of privacy. These tensions, in conjunction with the increase in deaths due to disease, eventually gave rise to heightened fears of sorcery (*vuloyi*) and encouraged an expansion of the definition of potential sources of sorcery from one that limited suspicion to close kin to one that included more distant relations and even strangers. This increased fear of sorcery may also have resulted from the fact that certain individuals, and particularly those who worked in the mines, were more prosperous than their kinsmen and neighbors who continued to rely on the rural economy. This prosperity—combined with the location of the mines in the forest, a region associated with pollution and dangerous forces of the bush—opened the returning mine workers to accusations of sorcery.[29]

Muhayirwa's tenure of office, from December 1932 to September 1935, therefore, corresponded to a period of social upheaval, economic decline, and an increase in sickness and death. It is consequently not surprising that Muhayirwa, as the instrument of Belgian administration, became a focus for local unrest and that attempts were made to have him removed as *grand chef.* This, however, is an external view of events, which may have been held by some Bashu, but not, I suggest, by the majority.

From the point of view of most Bashu informants who lived through this period, the call for Muhayirwa's removal and his replacement by Vyogho's son Muhashu was not simply an attempt to replace an unpopular chief. Rather, it was a means of restoring the well-being of the land by reestablishing ritual control, and it represented a reassertion of Bashu ideas about the nature and function of chiefship. To understand this, it is necessary to examine how the Bashu themselves viewed the changes that were occurring in their society during the early years of the colonial period. This view is represented in the following exchange between myself and an elderly informant, who had witnessed these changes:

Interviewer: Why did the Bashu call for Muhashu's return?
Informant: We had no *mwami* then and the land was uncov-
                 ered. There were many deaths. Our gardens were
                 spoiled by wild animals and there were many sor-
                 cerers. Finally the locusts came and destroyed the
                 crops. That is why we called for Muhashu.
Interviewer: But you had a chief in Muhayirwa.
Informant: Muhayirwa was not a true *mwami*. He had no
                 mbita and was just an overseer.
Interviewer: But Muhashu had no *mbita* at that time either.
Informant: Yes, but he was the son of Vyogho and had the
                 power.[30]

In response to a similar question about Muhayirwa's overthrow, another informant described the following incident, which occurred during Muhayirwa's rule in 1933:

There was a rainmaker from Mbulamasi named Swera, who brought down a hailstorm on the land of a neighboring chief, Hangi, whose people Swera had previously assisted by bringing rain for their crops. The reason for this attack was Hangi's refusal to give Swera some beer [symbolic of his refusal to recognize his association with Swera]. This angered Swera, who brought down the hail and destroyed the area's crops.[31]

This event, according to the informant, had occurred because there was no *mwami* to control the use of ritual power over the land and to prevent individuals from using their ritual powers for individual ends.

For these informants, declining productivity of the land, the destruction of crops by locusts, wild animals and hailstorms, and increases in sorcery, sickness, and death, were a product of Belgian interference in the Bashu political system, the removal of a central chief, and the consequent disruption of ritual control over the land. To restore the well-being of the land, it was necessary to reestablish ritual control. This required the investiture of a new *mwami w'embita*.

This is not to suggest that the Bashu were unable to understand the economic developments that were changing their society. Bashu informants give clear indication that they understood that their declining productivity was related to increased taxation, demands for corvée labor, the loss of markets, and decreases in the prices paid for their produce, even if the wider forces of capitalist development behind these conditions were imperceptible. Moreover, it is equally apparent that the Bashu did not see the renewal of chiefship as the only solution to their problems. Other actions were taken to improve conditions, including the passive resistance campaign mentioned above, the abandonment of consolidated villages—a problem that plagued Belgian administrators throughout the 1930s—and migration away from administrative centers and, in some cases, to Uganda. Yet their understanding of economic forces did not explain the relationship between Belgian rule and increases in sickness, death, and the use of sorcery, or the ravages of wild animals, hailstorms, and locusts. Nor did it provide a solution for these problems. These occurrences called for other explanations and actions, for they were symptomatic of the failure to regulate properly the relationship between the world of the homestead and the forces of the bush, and every Bashu knew that this failure was due to the absence of a central ritual chief and could only be remedied by the return of Vyogho's successor, Muhashu. It is possible, moreover, that despite their understanding of how certain elements of the colonial economy were causing a decline in their productivity, many Bashu saw Muhashu's return as a panacea for all their ills, and not just for those that were directly attributable to forces of the bush. The *mwami* was, after all, seen as able to insure the productivity of the land.

The choice of Muhashu rather than Buanga to succeed Vyogho appears to have been made because Muhashu had established a wider base of ritual support in Isale among Vyogho's former *basingya*. It thus re-

flects the continued importance of this support. Muhashu formed marriage alliances with the Bahera of Ngukwe, the Batangi of Ihera (who
had served as Vyogho's *baghula* and therefore possessed his lower jawbone
and the regalia necessary for Muhashu's investiture), and the descendants
of Kahindolya, who were the original chiefs of Vusekuli. Muhashu evidently also gained the support of the Bashu of Mbulamasi; the rainchief
Mutunzi; the Babito chiefs of Maseki; Patanguli, grandson of Muhoyo;
and Murondoro, the grandson of Mutsora, who became Muhashu's
mukulu. While Buanga succeeded in gaining some support in northern
Isale and eventually used it to undermine Muhashu's authority in the area,
his support was not as extensive or widespread as Muhashu's at the time
of their return from exile.[32]

The call for Muhashu's investiture and the removal of Muhayirwa was
not, therefore, an attempt to undermine central authority, as Belgian
administrators thought. It was, to the contrary, an effort to reestablish
central authority, but within the context of Bashu rather than Belgian
political perceptions about the nature and function of chiefship. Moreover, the choice of Muhashu rather than Buanga reflects the continuation
of previous patterns of political competition based on Bashu political

TABLE 8
THE ALLIES OF MUHASHU (1935)

Allies	Status	Office
Bahera of Ngukwe	Senior Bahera settlement	*semombo*
Batangi of Ihera (formerly Lulinda)	royal ritualists for early Babito chiefs	*baghula*
Descendants of Kahindolya	earliest chiefs in Isale	*bakaka*
Bashu of Mbulamasi	diviners of the land, rainmakers	*semombo*
Babito of Maseki	chiefs of Maseki	—
Babito of Vusekuli (Morondoro, grand-son of Mutsora)	chiefs of Vusekuli	*mukulu*
Mutunzi (Vusali)	rainchief	*musingya*, provided *mbita*
Patanguli (Mutendero)	Babito chief, grandson of Muhoyo; included among subjects were Baswaga of Katikale and Bashu of Manighi	—

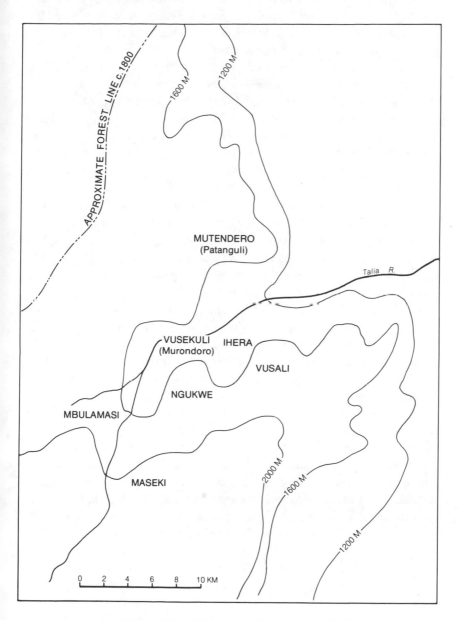

MAP 7. The Alliances of Muhashu, ca. 1935

values. With Muhashu's investiture as *grand chef* and his subsequent accession to the position of *mwami w'embita,* the Bashu succeeded in temporarily reasserting their own political values by subordinating the office of
grand chef to the politics of ritual chiefship.

Yet the pattern of political competition and the nature of chiefship had
clearly been altered by the Belgian presence. Encounters between rival
chiefs were no longer determined by popular support and ritual legitimacy alone. The encapsulation of the Bashu political system by a colonial
polity required that popular support be expressed in terms that conformed to Belgian notions of political legitimacy. Thus, the Bashu attacked Muhayirwa for abusing his secular authority and for corruption
rather than for not possessing ritual power. They may also have presented traditions supporting Muhashu's legitimacy. For the official history of the Bashu chiefdoms that appears in Belgian administrative
reports after 1935 once more attests to the legitimacy of Vyogho's
descendants and reduces Muhayirwa's line to the status of subchiefs,
though again this alteration may have been produced by the Belgians
themselves without Bashu assistance.[33]

The encapsulation process also altered the nature of support-building.
Before the colonial period, the Babito sought support among the various
commoner lineages of the region, with special attention being given to
those lineages that possessed ritual influence over the land. Fellow Babito
chiefs were generally viewed as potential rivals rather than as allies,
although, as we have seen, close ritual cooperation did occur between
certain Babito chiefs. During the colonial period, rivals for chiefship
began to pay greater attention to acquiring the support of fellow Babito
chiefs as well as important commoner lineages. Thus, as noted above,
Muhashu sought the support of the Babito chiefs of Maseki and of
Patanguli, the grandson of Muhoyo, as well as that of his family's traditional Babito allies, the descendants of Mutsora in Vusekuli. The growing
importance of Babito support can be explained partly in terms of traditional support-building patterns. By the beginning of the twentieth century, several lines of Babito chiefs had succeeded in establishing
independent spheres of ritual influence over parts of Isale and thus in
acquiring an important role in the ritual maintenance of these areas. Like
former non-Babito chiefs, their cooperation and support were thus ritually and politically important and had to be acquired by Vyogho and his
successors. On the other hand, the increased importance of Babito support can also be seen as an accommodation to Belgian perceptions of
political legitimacy. The Belgians often described the Bashu chiefdoms

as conquest states, which had been established by the Babito through force of arms. While there appears to be little basis in fact for this belief, it affected the way in which Belgian officials viewed Bashu politics. Following this view, succession questions were seen as the exclusive concern of the Babito themselves. Belgian officials therefore attached considerable importance to the preferences expressed by Babito subchiefs in making their choice of a new *grand chef*. Consequently, Muhashu and his later rival Buanga had to acquire the support of these other Babito leaders in order to win recognition in the eyes of Belgian administrators. The nature of support-building was thus affected by the Belgian presence.

The nature of chiefship was also altered by the encapsulation process. While Muhashu possessed both ritual and secular authority and thus resembled Vyogho in terms of his political power, his secular powers and responsibilities were much more extensive than those of his father. Moreover, the hierarchy of subchiefs recognized by the Belgians was more widespread and closely integrated than Vyogho's hierarchy. With the Belgians, every part of the Bashu region fell within the local administrative network, and thus under Muhashu's authority. Finally, Muhashu's administrative responsibilities required him to travel much more extensively than his predecessor. Invested chiefs had rarely left their courts prior to the establishment of Belgian rule.

These changes conflicted with Bashu conceptions and expectations about the ideal behavior and function of a *mwami w'embita* and opened Muhashu to local criticism. To reduce this criticism, he tried to minimize his role as *grand chef*, restricting his activities primarily to performing ritual functions and leaving local administration to his subchiefs. This strategy, however, angered the Belgians, who attacked him for not carrying out his administrative duties. In the end he was unable to resolve the contradictions inherent in his dual role as *grand chef* and *mwami w'embita*. The Belgians accused him of having agreed to keep a low profile in order to gain the support of other chiefs for his candidacy for *grand chef*,[34] while the Bashu began to associate the continuing deterioration of the land and society with his nontraditional behavior. This charge seriously undermined his ritual credibility and popular support and ultimately led the people of northern Isale to reject his authority.

By the 1940s the Bashu region was firmly enmeshed in the colonial economy and was, from a Belgian viewpoint, prospering. Arabica coffee had been successfully introduced to offset the decline in Bashu income created by the cessation of food purchases by the Minière des Grandes Lacs (Joset, 1939: 22). Thirty thousand tons of coffee were marketed in

1943. Wheat, which had been introduced earlier in the higher regions, was doing well, and soybeans were introduced in Isale.[35] In addition, the MGL recommenced purchasing Bashu crops in 1939.[36] Cassava production was encouraged as a substitute for eleusine because of its resistance to climatic change and in response to MGL needs, while potatoes were introduced in the higher regions as a new source of starch. Labor recruitment to the mines and European plantations also increased steadily throughout the 1940s, resulting in the employment of over a third of the Bashu male population by 1945.[37]

Despite these indicators of economic development, Bashu testimonies and sporadic references amidst the generally positive reports of local Belgian administrators provide a somewhat more negative view of what these changes meant in terms of the material well-being of Bashu society.

To begin with, the new crops were a mixed blessing. While coffee provided new sources of income and could be grown in conjunction with existing subsistence crops without greatly disrupting normal agricultural activities, its production did require inputs of labor and land resources that, in conjunction with increased pressure on male labor, limited the production of food crops. Soybean production was a greater burden, for soybeans had to be cultivated on the same land and during the same period as beans, which were a primary source of protein in the Bashu diet. Because the people of Isale were obligated to grow soybeans, and because they refused to include them in their own diet, their supply of protein was reduced.[38] The gradual substitution of cassava for eleusine further reduced protein supplies. In the higher elevations to the south, increased dependence on potatoes created a serious problem when in 1943 the potatoes, along with wheat, were hit by mildew, creating a famine.[39]

In 1936 all settlements below 1500 meters were evacuated as a further precaution against the spread of sleeping sickness.[40] This move, like the earlier restriction of settlements below 1000 meters, forced the Bashu to abandon gardens and further reduced land available for cultivation. Ironically, these lower slopes were the areas most suitable for cassava production, which, in other ways, the Belgians were trying to encourage. Forcing people to move above 1500 meters also increased population densities and pressure for land above this altitude. According to a Belgian report, overcrowding had already begun to occur in the Maseki region by 1935.[41] Thus, restrictions on settlement below 1500 meters created a real problem in this area. Land pressures were further increased by the beginning of European settlement and the alienation of lands for European coffee plantations. This process did not proceed very far, however, until the end of the Second World War. Nonetheless, economic conditions within the Bashu chiefdoms were not as good as Belgian reports indicate. While conditions appear to have been better than in the early 1930s, economic advancement within the rural economy was restricted, which may account for the large numbers of Bashu men who chose to leave the local peasant economy to work in the mines or on European plantations.[42]

Moreover, as in the 1930s, economic conditions were not the only problems facing the Bashu during this period. By the early 1940s, plague was becoming a recurrent feature of village life in the mountains and was the cause of numerous deaths. In addition, increased demands on female labor and the need for women to fill many of the roles vacated by their absent husbands appear to have led to demands by women for a greater share in whatever benefits accrued from their labor. This in turn contributed to increased social tensions between men and women and to the spread of new fears about female witches called *bambakali*. While accusations against women who were said to be *bambakali* seem to have served to reassert the dominance of men in Bashu society by providing a justification for punishing women who were seen as too demanding, the belief itself was a logical outgrowth of Bashu ideas concerning the ambivalent role of women. *Bambakali* were thus defined as women who had rejected their responsibility to the homestead in favor of their associations with the world of the bush. Their intrusion into Bashu society was therefore symptomatic of the breakdown in the regulation of the relationship between the homestead and the bush, interpreted in terms of conventional Bashu cosmology.[43]

Finally, 1943 and 1944 saw extensive drought throughout the Bashu region, which exacerbated the famine in the south and initiated famine in the north.[44] This ecological disaster, combined with increases in disease and the emergence of *bambakali*, was for many Bashu living in northern Isale a direct result of Muhashu's failure to provide adequate ritual leadership, which in turn had resulted from his failure to conform to standards of behavior appropriate to a *mwami w'embita*. Accordingly, the people of this region turned to Muhashu's older brother, Buanga, and invested him as *mwami* at the end of 1945. Muhashu subsequently moved his settlement to southern Isale, where his support remained strong, and Isale was once again divided between northern and southern factions.[45]

Why did the people of northern Isale ascribe the famine to Muhashu's shortcomings and invest his brother as *mwami?* The answer appears to be that Buanga succeeded in acquiring ritual support in northern Isale prior to the onset of the famine and that when the famine occurred, the people of the region felt that Buanga was in a better position to renew the land. Buanga's investiture, therefore, represents a continuation of the politics of ritual chiefship within Isale.

There is clear evidence that Buanga was angered over the *basingya's* decision to invest Muhashu rather than himself. This anger was increased by the fact that the position of *mukulu*, which, as firstborn son, he might have inherited, had become hereditary among Mutsora's descendants. Buanga was thus left without substantial political authority in Isale. He is, in fact, said to have gone to the shrine at Vyogho's grave following Muhashu's accession and desecrated it by exposing its contents to the sunlight. According to one informant, he accompanied this action with the lament, "I who was a distributor am now a receiver."[46]

Buanga evidently set about deliberately undermining Muhashu's authority in northern Isale following Muhashu's investiture, though his search for ritual support may have begun earlier. He subsequently formed alliances with the Babito chiefs Patanguli, who had previously supported Muhashu; Muhayirwa; and a branch of Mutsora's family, who provided his *mukulu* as well as a link to the ritually important region of Vusekuli. Buanga also formed marriage alliances with several important non-Babito groups, including the descendants of Kaherataba, who were the 'clearers of the forest' at Mughulungu; the Basumba of Makungwe; the Bashu of Kalambi, who were 'clearers of the forest' and guardians of Luvango's tomb; and the Baswaga of Katikale. He formed an additional alliance with the Batangi of Lulinda, whose relatives in Ihera had served as Vyogho's *baghula* and continued to support Muhashu's chiefship. This alliance allowed Buanga to claim the support of this ancient family of royal ritualists. Together, these alliances established Buanga's claim to chiefship in northern Isale and allowed him to discredit Muhashu during the drought of 1944. As in 1935 political events among the Bashu were shaped by the politics of ritual chiefship.

Buanga's accession was a serious blow to Muhashu's chiefship and may have affected his health, for he died shortly thereafter. Muhashu's death convinced many people in Isale that he had been a false *mwami* and that Buanga should have been invested earlier instead of Muhashu. Buanga therefore assumed the role of successor to Vyogho, although his authority was opposed by a few of Muhashu's closest supporters in southern Isale. Buanga did not, however, inherit the office of *grand chef,* for while there were those who thought that he should, the Belgians felt that he was too old. Moreover, he is said to have been partially deaf, which made it

TABLE 9
THE ALLIES OF BUANGA (1945)

Allies	Status	Office
Muhayirwa (Makungwe)	Babito chiefs; descendants of Visalu and Molero	—
Patanguli (Mutendero)	Babito chief; subjects include Baswaga of Katikale and Bashu of Manighi	—
Kakonda (Vusekuli)	grandson of Mutsora; claimed right to control Vusekuli	*mukulu*
Descendants of Kaherataba (Mughulungu)	clearers of the forest	*basingya*, provided *mbita*
Batangi of Lulinda	royal ritualists for early Babito chiefs	*baghula*
Basumba of Makungwe	first occupants of the land	keepers of royal fire
Baswaga of Katikale	clearers of the forest	*semombo*

difficult for the Belgians to communicate with him.[47] In addition, there was strong Bashu opposition to his investiture as *grand chef*.

The Belgians consequently convened a meeting of Bashu notables on March 28, 1945, to discuss the succession and choose a *grand chef*. The Bashu notables present at this meeting appear to have initially interpreted its purpose as choosing a successor to Muhashu's ritual chiefship. This question was difficult enough, given the split between Buanga's supporters and those of Muhashu's son Kaghoma. However, when the Belgians indicated that the issue was not Muhashu's ritual successor but the investiture of a new paramount chief, a new *grand chef*, the decision became even more difficult. It took two years to reach a decision. This delay was caused by competition between supporters of Buanga and supporters of Kaghoma and by the refusal of chiefs in Maseki and Malio to recognize either man. In the end, when the Belgians made it clear that a decision had to be made, the Bashu chose to reinstall Muhayirwa.[48]

Given the disaster that Muhayirwa's previous investiture as *grand chef* had produced, his reappointment is somewhat curious. However, the situation had changed. Previously, Vyogho's successors had been in exile. From a Bashu perspective this meant that the land was uncovered. Now

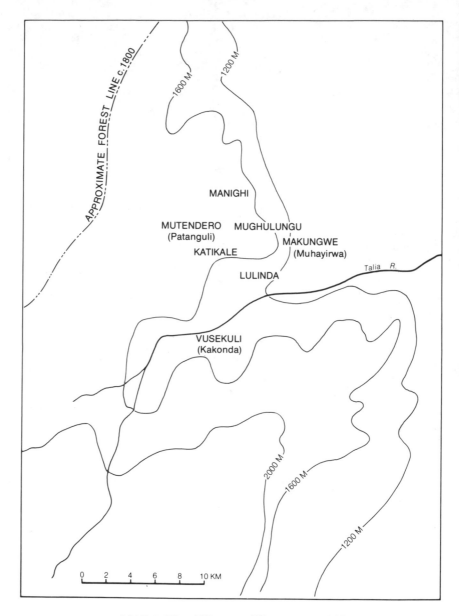

MAP 8. The Alliances of Buanga, ca. 1945

the Bashu had a legitimate ritual chief in Buanga, even though his authority in southern Isale was challenged by Muhashu's son Kaghoma. Also, in contrast to before, Muhayirwa is said to have agreed not to challenge Buanga's ritual authority as he had previously done by blocking his and Muhashu's return to Isale. Finally, the Bashu may have seen Muhayirwa's investiture as *grand chef* as a way of avoiding the disasters that in many peoples' eyes, had resulted from Muhashu's having combined both the office of *grand chef* and of *mwami w'embita* in one person.

With Muhayirwa's appointment, ritual and secular authority were once more divided and have remained so to the present day. Muhayirwa was succeeded by his son Muchozi, who, like his father, has shown no inclination to serve as a ritual leader among the Bashu. As for ritual leadership in Isale, Buanga died in 1950 and there have been no customary investitures in his family since then. Moreover, Muhashu's son Kaghoma has yet to be invested as *mwami*. There is thus no ritual chief over the whole Bashu region or even over Isale. As noted in the introduction, the failure to invest a new *mwami w'embita* has been a source of considerable concern among Bashu elders, many of whom continue to accept the validity of the relationship between chiefship and the problem of environmental control. There is evidence, however, that both Kasayi, the son of Buanga, and Kaghoma have over the last decade slowly built up support bases in their respective regions of Isale, and that a major famine may eventually lead to the investiture of one or both men.

CONCLUSION

Belgian attempts to incorporate Bashu chiefship into their colonial administration initiated a conflict between Belgian administrators, who saw chiefs as local administrators, and Bashu notables, who viewed them as ritual intermediaries regulating the relationship between the world of the homestead and the forces of the bush. As we have seen, these two perspectives ultimately were incompatible. Neither side could accept the other's definition of what a chief should be, and the chiefs, caught in the middle of this debate, were unable to fulfill the expectations of either side. In the end, the conflict was resolved by separating Bashu chiefship into its political and ritual components. This compromise provided the Belgians with administrative chiefs, while it permitted the Bashu to maintain the crucial ritual functions of chiefship.

Competition between Belgian authorities and Bashu notables over the definition of chiefship, like the conflict between advocates of political violence and the followers of the politics of ritual chiefship during the

1890s, attests to the persistence of Bashu political ideas and values in the face of rapid and deep-seated social change. This persistence may reflect Bashu conservatism and a desire to retain traditional values. Yet the Bashu have exhibited a general willingness to adopt new fashions, architectural styles, foods, ritual practices, and even ideas about misfortune. Why, then, have Bashu ideas about the nature and function of chiefship been so resistant to change?

One reason may be the self-sustaining nature of Bashu political ideas. Like the Zande concepts of witchcraft and magic described by Evans-Pritchard (1937: 320), Bashu ideas about the nature and function of chiefship " . . . are eminently coherent, being interrelated by a network of logical ties, and . . . so ordered that they never too crudely contradict sensory experience, but instead, experience seems to justify them." Thus we have seen that while the occurrence of plenty or famine does not always coincide with cycles of chiefship, and famines may occur during the reign of a strong chief, this incongruity never seems to challenge seriously the validity of Bashu ideas concerning the association of strong central ritual control and chiefly power with conditions of plenty. For such anomalies can be explained in terms of Bashu ideas about chiefship as resulting from the anger of the *mwami*, or the failure of a recalcitrant ritual leader.

A second explanation for the persistence of Bashu ideas about the nature and function of chiefship may lie in the central position of chiefship within the Bashu view of the world in which they live. As indicated in chapter one, Bashu chiefship is intricately tied to their wider view of nature and society. It is a logical and necessary corollary to their view of the universe as a place that is divided between spheres of homestead and bush, nature and culture, and of their assumption that good fortune, plenty, and the regular cyclical passage of ecological time depend on the coordinated activities of ritual specialists who mediate between these two spheres. In other words, Bashu chiefship rests upon, and is part of, a comprehensive view of the universe. This close integration of Bashu political values with their wider view of the world makes these values resistant to change and suggests that, like the demonology of sixteenth-century European scholasticists, Bashu ideas about the nature and function of chiefship will continue to grasp popular imagination until the wider underlying worldview upon which they rest is abandoned or transformed (Trevor-Roper, 1969). All of this is not to say that Bashu political ideas are static. Changes have occurred in the way in which the Bashu view chiefship. Yet these changes have evolved slowly and represent gradual redefinitions rather than radical transformations of chiefship.

CONCLUSION:

CONTINUITY AND CHANGE IN THE POLITICS OF RITUAL CHIEFSHIP

Bashu political experience during the nineteenth and early twentieth centuries resembled that of a number of other societies in eastern Africa, and was, in fact, shaped by similar economic and social forces. Yet I have argued that the specific character of the Bashu experience was defined to a remarkable degree by Bashu perceptions of the world in which they lived and of the critical role played by chiefship in the maintenance of that world.

Bashu politics, like Bashu chiefship, centered on the fundamental problem of regulating the relationship between nature and culture, between the world of the homestead and the world of the bush. Chiefs were responsible for maintaining the well-being of society. Yet, as we have seen, they depended on the cooperation of local ritual leaders and former chiefs. For without this support, the forces of nature could not be domesticated and the productivity of the land and society insured. Achieving and maintaining the support of local ritualists—expanding the temporal dimension of chiefship—was thus a major concern of every chief and candidate for chiefship and was a primary goal of political competition, toward which political actors dedicated their economic, social, and occasionally military resources. The overriding importance of this goal was perhaps best exemplified by the actions of Kasumbakali, who, as we have seen, possessed military resources that permitted him to intimidate the people of Isale into supporting him at the end of the nineteenth century. Yet he still sought the support of local ritual leaders. Despite his military resources, his political behavior conformed to Bashu notions of legitimacy.

Success in gaining the support of ritual leaders was dependent on a chief's political skills, the influence of his maternal uncles, and, presumably, his access to wealth. Differences in wealth, however, are sometimes difficult to discern from available evidence. The initial success of the Babito in establishing their authority in Isale was no doubt related to their superior access to livestock and possibly salt. So too, Luvango's overthrow of Visalu and Mutsora's defeat of Ghotya were evidently based on

Luvango's and Mutsora's superior economic resources. Yet, if we look at Kivoto's competition with Muhoyo, or Vyogho's conflict with Kasumbakali, it is more difficult to distinguish victors from losers on economic grounds. While it is possible that further research into Bashu economic history may provide a clearer picture of the economics of ritual chiefship and a fuller explanation for the relative success of various competitors for chiefship, economics may not always have been the primary factor in determining political success. Diplomacy, personality, and the influences of one's maternal relations may have played equally important roles in particular competitions.

The need to incorporate local ritual leaders and former chiefs into the structures of chiefship had two important consequences for the nature of political relations among the Bashu. First, because local ritual support was a finite resource and thus an object of competition, local leaders were able to maintain a certain degree of autonomy and exact a price for their cooperation. This autonomy introduced an element of instability into Bashu chiefdoms and inhibited centralization, for it permitted local leaders to shift support from one chief to another. To counteract this centrifugal tendency chiefs attempted to strengthen their ties with local leaders by providing them with a vested interest in their chiefship. The office of *mombo* adopted by Mukunyu and his descendants accomplished this by offering every marriage ally a chance to become maternal uncles to the next *mwami*. The practice of dividing the office of *semombo*, father of the *mombo*, between a number of different allies, instituted by Kivoto and Vyogho, served a similar function by increasing the number and groups that were "maternal uncles" to the future *mwami*. The distribution of other titles and offices further strengthened a chief's ties to his ritual supporters and counteracted centrifugal tendencies.

The need to coordinate ritual authority over the land, and thus to incorporate local ritual leaders and former chiefs into the structures of chiefship, may also have contributed to the definition of political boundaries among the Bashu. Differences in the distribution of crops, the timing of agricultural cycles, and the occurrence of plenty and famine within the Bashu region inhibited ritual cooperation, and thus political consolidation, beyond the boundaries of ecologically defined zones. While these limits were not insurmountable, as seen in the extension of Vyogho's authority beyond Isale at the end of the nineteenth century, they appear to have contributed to the definition of the arenas within which political actors vied for support. They also served to help define the maximal boundaries of Bashu chiefdoms during most of the nineteenth century.

Vyogho's success in superseding these limits was made possible by major changes in the economic and social environment of the Bashu region and by the onset of large-scale natural disasters. By the middle years of the twentieth century, different conditions prevailed, and there was a movement back to regional autonomy following the death of Buanga in 1950. The importance of consolidating ritual control over the land within Isale also established Vusekuli, located in the center of Isale and an important political bridge between the northern and southern halves of the region, as a focus for competition among the Babito chiefs of Isale.

While rivals for chiefship continually vied for the support of local ritual leaders, their relative success was periodically tested and defined by the onset of adverse environmental conditions: extended rains, drought, locusts, or external raids. When disasters of this type occurred, it was up to a chief to explain why, since these events challenged his credibility and legitimacy. These explanations took a variety of forms. Thus Mutsora apparently ascribed the famine that followed Luvango's death to Ghotya's refusal to accept his authority over Vusekuli and to the disruption of ritual control over the land that this refusal caused. In the Mughulungu region Molero attributed the same famine to Luvango's earlier usurpation of Visalu's authority. Similarly, the combination of disasters that hit Isale at the end of the nineteenth century was attributed by Vyogho and his supporters to Kasumbakali's total disruption of ritual control over the land. Still later, in the 1940s, the people of northern Isale were convinced that the drought that struck their lands was the product of Muhashu's unorthodox behavior. Other explanations were presumably provided by less successful chiefs. However, Bashu traditions tend to incorporate only those explanations which are ultimately accepted by the people. It is thus difficult to know what form these other explanations took.

Successful explanations were consistent with Bashu ideas concerning the ritual maintenance of the land. Moreover, by placing ultimate responsibility for the disaster on a rival chief or recalcitrant local leader, each explanation accounted for the disaster without undermining the ritual credibility of the chief who provided the explanation. Each explanation also suggested a solution that, if accepted, would ultimately strengthen the position of the chief who provided it. Thus, Mutsora's explanation called for Ghotya's acceptance of his authority in Vusekuli before the famine would end. Molero's explanation required that he be invested as chief over the Mughulungu area. Vyogho's explanation required that Kasumbakali be driven from Isale, and Buanga's called for the people of Isale to accept him rather than Muhashu as their chief.

The congruence between a chief's explanation for famine and his political interests does not necessarily indicate that competitors manipulated
popular beliefs for their own political ends. Designating a rival chief's
opposition as having caused a famine, after all, was consistent with Bashu
ideas concerning the causes of famine. The possibility of manipulation
cannot, however, be ruled out. To begin with, belief does not preclude
the manipulation of belief. Thus, a man may believe in the power of
sorcerers but still manipulate the belief to attack a rival. So too, a chief
may have shared Bashu ideas about the causes of famine and yet have
accused a rival of having caused the famine by his conflicting use of ritual
power in order to gain the rival's submission. If the submission did not
end the famine, the chief could seek another cause afterward or he might
even seek a different cause simultaneously. It is also possible that some
political competitors did not share popular beliefs and simply conformed
to them in order to gain legitimacy. The introduction of an alternative
political ideology during the colonial period may have increased this
possibility. In the end, therefore, it is perhaps impossible to determine the
degree to which a given political actor conformed to Bashu political ideas
because he shared them or because he needed to acquire legitimacy, and
thus the degree to which popular ideas were manipulated. All that can
be said is that either way, Bashu political ideas shaped patterns of political
action.

Providing an explanation that conformed to local belief structures and
deflected responsibility for a disaster did not, of course, insure its acceptance, especially given the existence of alternative explanations. A chief's
ability to maintain his own explanation and to have it accepted depended
in large part on his success in winning the support of local ritual leaders
and former chiefs, since no chief could claim to regulate the land without
this local support. Thus, a chief who had been successful in accumulating
local ritual support would normally be able to maintain his explanation
until the famine ended, while a chief who failed to gain local support
would be unable to do so and would eventually have to recognize the
authority of a more successful chief or leave the arena of competition,
as Kivoto chose to do following his defeat at Mutendero.

Famines can thus be seen to have created encounters between rivals for
chiefship and to have defined the actual distribution of ritual and thus
political authority among the Bashu. The impact of natural disasters on
Bashu politics was therefore not random, as it often appears to be in the
histories of other African societies, but was in large measure predeter-
mined by the patterns of support building and subversion that preceded

them. Famines were in a sense an integral part of Bashu politics, although from a Bashu viewpoint it would be more accurate to state that politics was a part of a wider system of ecological control, since the Bashu were less concerned with how natural disasters affected politics than with how politics and political competition affected the occurrence of famine and plenty.

Finally, Bashu perceptions of politics and of its relationship to the changing conditions of the world in which they lived provided a conceptual framework through which the Bashu interpreted major changes in their social and economic environment and responded to them. Thus, I have argued that the social and political disruption that resulted from the spread of long-distance trade and firearms at the end of the nineteenth century was interpreted by the Bashu as a manifestation of the interruption of ritual control over the land, which the expansion of trade and firearms had indirectly caused, and that the Bashu responded accordingly by attempting to reconsolidate ritual control over the land. Cosmology served a similar function in explaining, and suggesting responses to, the social and economic disruption caused by the establishment of Belgian colonial rule and the incorporation of the Bashu region into the periphery of capitalist development during the early years of the twentieth century.

Bashu conceptions of chiefship can thus be seen to have shaped the ways in which the Bashu thought about politics and to have colored their political actions accordingly. Ideology was in a sense an underlying structure of Bashu political history. Yet Bashu political ideas were by no means immune to the historical forces that swelled around them. The political actions that they shaped and changes in the social and economic environment of the Bashu region modified and refined the ways in which the Bashu viewed chiefship.

In chapter two I argued that Bashu chiefship on the eve of Babito political expansion into the Mitumbas was based primarily on the possession of specialized ritual powers over the rain, and that the power of rainchiefs was thus conceptually equivalent to that of other ritual leaders. I also suggested that rainchiefs emerged as firsts among equals in their relations with other ritual leaders and became a focus for the coordination of ritual influence over the land within relatively small localized areas by the beginning of the nineteenth century. This transformation resulted from the central importance of rain control and from the superior economic resources that were consequently available to rainchiefs. As a result of their newfound significance, rainchiefs came to be identified with the general well-being of the land.

By the end of the nineteenth century Bashu chiefship had undergone an ideological transformation. Chiefs were no longer regarded as ritual specialists and possessed no special ritual power over the land. They were no longer rainchiefs. Instead, their role as the central coordinator of ritual had become the primary conceptual basis for their political authority. This shift in emphasis was accompanied by an elaboration of the ritual functions attached to the role of ritual coordinator and an increase in the *mwami's* identification with the land. Vyogho was not simply a coordinator of ritual power; he was the central distributor of ritual power, controlling the distribution of ritual blessings, symbolized by his control of the *mikene*, dictating the timing of planting and harvest ceremonies, and controlling the use of rainmagic. Moreover, he not only coordinated ceremonies designed to cleanse the land but, through his ritual actions, also took on the "filth of the nation," thereby drawing together and completing the cleansing process. The *mwami's* position had thus become unique and essential to the welfare of his subjects. Without the *mwami*, the relationship between culture and nature could not be mediated. The well-being of the land was thus closely tied to the condition of the *mwami*, which was closely guarded. In addition, the geographical scale of chiefship had expanded from that of earlier rainchiefs, whose influence was limited to several adjacent ridges. Vyogho's chiefship incorporated all of Isale and parts of neighboring regions.

This transformation in Bashu ideas concerning the nature and scale of chiefship may represent the introduction of new ideas concerning the role of chiefs by the Babito. Bashu chiefs, after all, came to resemble their political counterparts among the Babito states in western Uganda in terms of their ritual functions and, to a certain degree, their political culture. Certain ideas may also have been acquired from the Batangi of Musindi at the time of Luvango's acquisition of ritual advisors from among the Batangi.

Yet, it is also possible that the transformation of Bashu chiefship was stimulated by changes in the social and economic environment of the Bashu region. Thus, the introduction of new economic resources in the hands of the Babito permitted the creation of larger chiefships. This in turn may have encouraged the movement away from specialized ritual authority. For as the Babito expanded their authority beyond the limits of existing chiefdoms, incorporating an increasing number of ritual specialists into their growing spheres of influence, their own specific ritual powers, whatever they may have been, may have become less important than their role as coordinators of ritual authority, especially if—as I have argued—their political success was tied to their ability to coordinate ritual activity over wide regions. If, on the other hand, the power of rainchiefs was conceptually limited to local areas because of subregional variations in the occurrence of storms—as suggested in chapter two—and not simply because rainchiefs lacked the economic resources needed to expand their authority beyond these regions, the shift away from specialized ritual leadership may have been a necessary precondition to any increase in the geographical limits of chiefship.

Of perhaps equal importance to the transformation of Bashu chiefship was the nature of political competition among the Bashu during the nineteenth century. This competition, culminating in the onset of major ecological disasters, reinforced the importance of ritual coordination. For the competition demonstrated that only a chief who could establish widespread ritual control over the land could maintain his explanation for the disaster and win credit for having ended it. In other words, Bashu notions about the importance of ritual coordination were reinforced each time a famine occurred. There was, therefore, a dialectical relationship between ideology and political competition through time, as Bashu ideas concerning the importance of ritual coordination shaped patterns of competition, which in turn reinforced Bashu ideology and encouraged the centralization of ritual authority.

Finally, we have seen that during the first half of the twentieth century Bashu chiefship was once more transformed as adjustments to the requirements of Belgian colonial rule led to a gradual separation of ritual and secular authority. The encapsulation of Bashu politics into the colonial world, moreover, modified patterns of political competition, as popular support had to be expressed in terms that were consistent with Belgian political values. Nonetheless, I suggest that these changes in the nature of chiefship and politics did not substantially alter the central position of the *mwami w'embita* in the Bashu world or eliminate the ritual content of Bashu politics. While new patterns of authority and competition have emerged within the Bashu region, the politics of ritual chiefship continues to operate alongside the bureaucratic politics of local administration, and occasionally to intersect with it. This interaction of ritual and modern bureaucratic politics has been at the heart of local political development in this corner of now independent Zaire for the last thirty years and should be the central focus of future studies on the modern political history of this region.

CONSTRUCTING A CHRONOLOGY
FOR BASHU HISTORY

The Bashu appear to have had little interest in affixing absolute dates to historical events prior to the establishment of colonial rule. Like many African peoples, their conceptions of time were functional and tied to their relationship with nature, the developmental cycles of social groups, and the social development of individuals. Time as a constant, evenly metered measurement of experience was foreign to their world view. For this reason, Bashu historical traditions are by themselves of little use in constructing absolute chronologies, though they do provide a relatively clear view of the order in which events occurred.

In order to construct a more precise picture of when particular events occurred it has been necessary to employ other data and methods. For the period following 1923 and the establishment of Belgian rule, the problem of constructing a chronology is relatively simple, for there is a wealth of archival data that can be combined with the sequential picture of events presented in Bashu traditions to give a detailed chronology of these events.

For the period between 1889 and 1923 the accounts of early European administrators and travellers, such as Stanley and Lugard, provide dates for events that occurred in the neighboring Semliki Valley region. These events, with their respective dates, can be tied into Bashu traditions at several points in order to date events within the Bashu region. These tie-ins provide a few hard dates as well as material for estimating the timing of other events.

For example, combining Bashu traditions with data from the Semliki Valley allows us to date within two or three years the death of the *mwami* Kivoto, who was a chief in Isale before the arrival of the Europeans. According to Bashu traditions, Kivoto was alive when the armies of the Nyoro king Kabarega began raiding the Isale region. The traditions also tell us that these raids continued for a short period after Kivoto's death. From Semliki sources we can date these raids to between the late 1880s and 1891, the date of Lugard's arrival at Katwe and his dispersal of the Nyoro forces. Combining these two sources of data we can tentatively date Kivoto's death at around 1890.

For the period prior to 1889, constructing a chronology becomes a more difficult task. There are no European records to tie into Bashu traditions, and very few other temporal guideposts exist that can be employed with equal certainty. Nonetheless, it is possible to construct an approximate chronology by employing notional generation averages. In table 10, I have constructed, following Oliver (1959) and Jones (1970), a chronology based on an average generation length of twenty-seven years, allowing a margin of error of plus or minus two years for every generation back from 1974, and an additional twenty years overall. Applying these dates to the genealogy of the ruling line of Bashu chiefs provides approximate dates for each chief. These dates correspond to the period of each chief's active political life, i.e., from the death of his father to his own death. This is not an average reign length, for the frequent occurrence of extended interregna among the Bashu often made the reign of a chief shorter than twenty-seven years. For the first four generations, these dates seem to be reasonably accurate. For example, we know from European records that Buanga died in 1950 and that his father, Vyogho, died in 1922, the year of the Ransbotijn expedition into the Bashu region. We can also conclude that Vyogho's father, Kivoto, died around 1890. While the death of Luvango is more difficult to date, we know from Bashu traditions that his death was followed by an extended period of wet weather, which caused a major famine in Isale. From climatological records (Nicholson, 1976: 147) it appears that the 1860s and 1870s were wetter than normal throughout the lake region of eastern Africa. If this wet spell corresponds to the rains that followed Luvango's death, we can tentatively date his death at around 1860. The dates of the last four chiefs, therefore, coincide with those of the proposed chronology based on notional generation averages, given the recommended margin of error. This correspondence suggests that the proposed dates of earlier generations may also be relatively accurate, although I lack the independent data needed to confirm this conclusion.

Assigning approximate dates to each chief allows us to date a wide range of events in Bashu history, for the traditions of both commoners and royals relate events to the reigns of specific chiefs. Generational averages, therefore, may provide a rough chronology for Bashu history prior to 1890. Readers are advised to consult Henige (1974), Cohen (1977), and Webster (1979) for recent discussions of some of the problems and possibilities of using generational data for establishing chronological relationships.

TABLE 10
TENTATIVE CHRONOLOGY FOR HISTORY OF ISALE

Generation	Proposed Dates	Senior Line of Babito Chiefs	Climatic Record	Events in Bushu	Events in Semliki
1	ca. 1974–1947 ±22	Kisuki			
2	ca. 1947–1920 ±24	Buanga, d. 1950	Drought, 1943–44	Belgian occupation, 1923	Sleeping sickness, 1905 Belgian posts in Semliki Valley, 1896, 1897
3	ca. 1920–1893 ±26	Vyogho, d. 1922	Drought ca. 1897	Famine, wars of Kasumbakali, Nyoro raids, 1890s	Banyoro at Katwe, 1889
4	ca. 1893–1866 ±28	Kivoto, d. ca. 1890	Heavy rains, 1860s, 1870s (Nicholson, 1976: 147)	Famine, succession struggle, fragmentation of Babito rule, 1860s	Toro establishes authority over Busongora
5	ca. 1866–1839 ±30	Luvango			
6	ca. 1839–1812 ±32	Mukunyu			Toro raids Busongora, 1830s, 1840s
7	ca. 1812–1785 ±34	Kavango	Sporadic drought conditions throughout 18th c. (Herring, 1979: 59–60)	Settlement of Babito chiefs in Isale	Expansion of Babito-Bahima rule west of Semliki, creation of chiefdoms of Bugaya and Kiyanja
8	ca. 1785–1758 ±36				
9	ca. 1758–1731 ±38			Settlement of Nande cultivators in Isale	
10	ca. 1731–1704 ±40		Major drought, 1720s (Herring, 1979: 59)	(9–11 gens.)	Consolidation of Babito-Bahima authority over Busongora, creation of Kingdom of Kisaka
11	ca. 1704–1677 ±42				
12	ca. 1677–1650 ±44				
13	ca. 1650–1623 ±46				
14	ca. 1623–1596 ±48		Major droughts, 1590–1620 (Herring, 1979: 57–58)	Settlement of Nande in Mitumbas following Baswaga traditions (13–15 gens.)	
15	ca. 1596–1569 ±50				

COMMONER LINEAGES AND FORMER CHIEFS

COMMONER LINEAGES

Bahera of Ngukwe — Senior branch of the Bahera clan, whose settlements are located throughout the southern half of the Bashu Collectivity. Luvango took a wife from this group, and they subsequently became the maternal uncles of his initial successor, Kivoto. They provided Kivoto and his son Vyogho with their *mombos*.

Bakira of Kirungwe — Initial allies of Luvango's firstborn son, Mutsora, in the Vusekuli region.

Banisanza of Lisasa — Formed an early alliance with Mukunyu and were maternal uncles to Mukunyu's initial successor, Visalu. Later made an alliance with Luvango and helped Luvango's grandson Kasumbakali in his attempt to gain control of Isale in the 1890s. Tsombira, the *mwami* of this group during the 1890s, was an ally of Karakwenzi.

Bashu of Manighi — *Bakonde* (clearers of the forest) of a large area in northern Isale. Originally directed the performance of royal rituals for the former *mwami* Mukirivuli. Made alliance with Luvango and served as his *bakaka* (royal buriers). Later supported Muhoyo in his successful usurpation of Kivoto's authority in northern Isale. Served as *baghula* (guardians of royal jawbone) for Muhoyo and his descendants.

Bashu of Mbulamasi — Important diviners of the land and rainmakers in southern Isale. Became *baghula* for Luvango and provided him with his *mombo*. Supported Kivoto against Muhoyo and became *semombo* of Kivoto and Vyogho.

Bashu of Mughulungu	(Descendants of Kaherataba) *Bakonde* of Mount Mughulungu. Gave land to Kavango when he first arrived in Isale. Later participated in Mukunyu's investiture as well as the investitures of Visalu and Molero. Shifted support to Luvango's descendants after succession struggle between Molero's sons and brother. Were maternal uncles of Vyogho's firstborn son, Mbuanga.
Bashu of Mwenye	Luvango won the support of this group by assisting them militarily. They later supported Muhoyo against Kivoto. Also supported Kasumbakali.
Basumba of Makungwe	First occupants of Mughulungu area. Became allies of Kavango and Mukunyu. Sided with Visalu in his dispute with Luvango and participated in Molero's investiture. Served as keepers of royal fire for Molero, Kivere, and Kamabo.
Baswaga of Katikale	*Bakonde* of region around Mutendero. Formed alliance with Mukunyu and were Luvango's maternal uncles. Supported Muhoyo against Kivoto.
Basyangwa of Muluka	Senior branch of the Basyangwa, who are said to be *bakonde* for most of the southern half of the Bashu collectivity. Allies of Kivoto and Vyogho. Maternal uncles of Kivoto's firstborn, Katsuba.
Batangi of Lulinda	Formed early alliance with Kavango and served as *bakaka* and directed the performance of royal rituals for Mukunyu, Visalu, and Molero. Later, following the succession war between Molero's sons and brother, they shifted their support to Kivoto and Vyogho. A branch of this family served as Vyogho's *baghula* and settled at Ihera.
Bito of Biabwe	*Bakumu b'emikene* and diviners of the land, formed marriage alliance with Luvango and subsequently became maternal uncles to Luvango's firstborn son, Mutsora. Mutsora's son Mukiritsa took a wife from this group.

Bito of Bunyuka

(Descendants of Sine) *bakonde* of Bunyuka. Important ally of Muhiyi, the former *mwami* of Bunyuka. Later formed marriage alliance with Luvango, whose firstborn son, Mutsora, took his first wife from this group.

FORMER BAMI

Baswaga of Ngulo

Senior line of chiefs among the Baswaga. Allied to Luvango. Later supported Muhoyo, to whom they were maternal uncles. Controlled the water oracle at Musalala.

Ghotya

Former *mwami* of the regions of Vusekuli and Vutungu. His position was usurped by Mutsora during the famine that followed Luvango's death. His family remained important ritual leaders in the region under the authority of Mutsora's descendants.

Muhiyi

Former chief of Bunyuka, who is generally credited by the Bashu as having discovered Isale. Kavango formed an early alliance with Muhiyi as a means of offsetting his dependence on the *mwami* Mukirivuli. Muhiyi's descendants later terminated this alliance when Kavango's grandson Luvango attempted to undermine their position by forming an alliance with one of their major supporters, the Bito of Bunyuka. Later, Luvango's son Muhoyo enlisted the support of Muhiyi's descendant Nunyu-Iremba in his conflict with Kivoto. Muhiyi's descendants have remained chiefs of Bunyuka while recognizing the present political authority of Kavango's descendants.

Mukirivuli

Former *mwami* of the regions surrounding Mount Mughulungu. Kavango paid tribute to Mukirivuli when he first arrived in Isale, and Mukirivuli invested Kavango's son, Mukunyu. Later, Luvango was able to terminate his family's dependence on Mukirivuli's family as a result of his success at alliance formation, his subversion of the authority of Mukirivuli's descendants, and a succession struggle between Mukirivuli's sons. Mukirivuli's elder son, Kisere, supported Visalu, while his younger son, Mbopi, made an alliance with Luvango. Mbopi subsequently supported Muhoyo against Kivoto.

Mukumbwa

Former Mubito *mwami* of Maseki region. His position was usurped by Luvango's son Kahese.

Mutunzi

Former *mwami* from Vusali, noted for his rainmaking skills. Formed an alliance with Vyogho and was the maternal uncle of Vyogho's successor, Muhashu.

NOTES

INTRODUCTION

1. This lake was renamed in 1972 in conjunction with changes in the political leadership of Uganda. Since it is likely to be renamed again in the near future, and because it is most widely known by this earlier, though regrettably colonial, designation, I have elected to use this name in the present study.

2. The distinction between Banande and Bakonjo appears to be based in large measure on the somewhat accidental placement of colonial boundaries in this region and the consequent separation of the mountain peoples living on the eastern slopes of the Ruwenzori Mountains in what is now Uganda, from their close relations on the western slopes of the Ruwenzoris and in the Mitumba Mountains of Zaire. In Uganda these people are known as Bakonjo, while in Zaire they are called Banande or, frequently, Wanande. The artificial nature of this division can be seen in the fact that the Bashu, who are classified as Nande, are in many ways closer linguistically and culturally to the Bakonjo than to peoples living in southern Bunande.

3. This sense of identification as "Bashu" by non-Bashu clans living within the Bashu chiefdoms may in part be a product of the colonial era: Belgian administrators used the term "Bashu" as a general classification for all of the people living within the Bashu-dominated chiefdoms. Another important Bashu settlement is located in Itala in southern Bunande. The historical relationships between Itala and Isale, however, remain unclear.

4. The Isale region has two rainy seasons separated by periods of drier weather. The short rains normally fall between the end of February and the beginning of May. May to August are drier months with some rains occurring primarily in the form of late afternoon showers. Heavier and longer rains occur between August and November followed by a dry season from the end of November through the middle of February. There is some regional variation in this general pattern with longer rains falling to the south in Maseki during the spring. This variation had important political consequences, as noted in chapter four.

5. See David W. Cohen, *Womunafu's Bunafu* (Princeton, 1977), for a particularly fascinating discussion of the role of political ideology in shaping action in relation to the movements of the Busoga Mukama figure Womunafu.

6. Steven Feierman's studies of Shambaa politics (1972, 1974) represent a notable exception to this general pattern. However, his decision to separate his material on Shambaa political history from his analysis of Shambaa political culture tends to obscure the historical relationship between these two aspects of Shambaa kingship.

7. Transcripts of these interviews, referred to in the following notes as Bashu Historical Texts (BHT), are in the author's possession and are available for consultation.

8. For a more complete analysis of the historical and cultural content of this myth, see R. Packard (1977 and 1980a).

9. The Bashu perform several types of songs. Narrative songs consist primarily of personal reflections and references to incidents in the singer's life. These are interspersed with traditional proverbs and occasional references to historical events.

10. The use of alliances as indicators of patterns of political interactions has been noted by a number of other historians, most notably Feierman (1972) and Cohen (1977).

CHAPTER ONE: FAMINE, PLENTY,
AND THE LIFE CYCLE OF *BWAMI*

1. Rapport de Reconnaissance, October 11–16, 1911, in Registre des Rapports Politiques-Kasindi, June 1906–June 1913, AZB.

2. These data are based on a government census taken in 1974.

3. These figures are based on monthly rainfall data collected by the meteorological recording station at ETSAV, Butembo (long. 29 17'; lat. .08 North; altitude 1747 m). Butembo is located 12 km west of Vuhovi, which is the administrative center of the Bashu region, and is subject to the same patterns of rainfall that affect Isale.

4. The breeding ground for these locusts is said to be the shores of Lake Mobutu (formerly Albert) to the north. Locust attacks occurred in recent times during the 1930s (see chapter eight) and threatened to do so again in November of 1974. The Bashu chiefs assembled in Isale to perform a major sacrifice designed to divert this potential disaster. To the gratification of all concerned, the locusts did not arrive.

5. For a more extensive discussion of Bashu concepts of misfortune and their relationship to wider cosmological ideas, see R. M. Packard (1980b).

6. The verb *-humbirya* is used to describe the action of beating sticks together to divert a hailstorm or of calming down someone who wishes to fight. It also describes actions taken by healers to transform substances taken from the bush into medicines for curing illnesses.

7. This information was provided by three Bashu informants who were themselves healers. A similar process is involved in a person's becoming an *mbandwa* medium. As elsewhere in the Lakes Plateau Region, *mbandwa* are possessed by spirits that give them the power of divination. Among the Bashu, however, the normal pantheon of Cwezi-related spirits does not, with the possible exception of Muhima, exist. Bashu *mbandwa* are possessed by local spirits of the bush.

8. See chapter two, pp. 68–69, for a discussion of the origins and use of this form of rainmagic.

9. See chapter seven, pp. 157–58, for a discussion of the origins of Nyavingi and its introduction among the Bashu.

10. This view of the ancestors contrasts with that described by Evans-Pritchard (1956) and others, in which ritual activity directed toward the spirits of the ancestors is designed to maintain or reestablish the separation between ancestors and living, and misfortune results from the interference of the ancestors in the world of the living. Bashu ancestors are more like elevated elders, and the term used to describe them is the same as that used to describe elders, *bakulu* or *basekulu*. For a discussion of ancestors as elders, see I. Kopytoff (1971) and James L. Brain (1973).

11. The verb *-tula* is used in other contexts to describe the action involved in breaking open a fruit, opening an envelope, and breaking into someone's house.

12. The significance of Bashu concepts of time for understanding chiefship only became apparent to me after completing my research in Zaire. The present discussion is therefore based on limited data and should be regarded as preliminary in nature.

13. Similar ideas about the collective ritual nature of political authority evidently occur among the Jukun of Nigeria, where, according to M. Young (1966: 147), "There was . . . a conception of the syncretic nature of kingship as an amalgam of competencies, of functional roles: king as ruler, as guardian of the corn, as rainmaker and even, since about 1800, as Mallam' to Islamized subjects." These elements were bestowed upon the king by priests in charge of each element at the king's installation and removed at his death. Similar conceptions can be seen closer at hand in the role of the *bwiru*, college of investors in precolonial Rwanda (Vansina, 1962; M. deHertefelt and A. Coupez, 1964), and in certain elements of Ganda kingship discussed by Ben Ray in his forthcoming study of Ganda kingship (personal communication).

14. The connection between giving milk and acknowledging political authority is seen in the following tradition, which describes the establishment of Bamate authority over the Bashu of Buhimba, located to the south of Isale.

> Shortly after the *mwami* Kihimba arrived in Buhimba, he was visited by two hunters who were servants of a neighboring *mwami* of the Bamate (a Nande subgroup). The hunters asked Kihimba for some food for their dogs. Kihimba gave them squash. The hunters then returned to their master and reported that they had found a powerful *mwami* living in the land which bordered on their master's chiefdom. The Mumate *mwami* told the hunters to return to this new *mwami* and ask him to give them some milk. The hunters followed these instructions but Kihimba did not wish to give them milk and told them that he had none. The hunters, however, saw drops of milk on Kihimba's beard and knew that he had lied. They therefore returned to their master and reported that Kihimba had refused to give them milk. This news enraged the Mumate *mwami* and caused him to send his hunters to Kihimba a third time, instructing them to kill Kihimba. The hunters followed their master's instructions and killed Kihimba with a spear. Following Kihimba's death the Bashu of Buhimba began paying tribute to the Bamate.

This tradition explains how the Bamate established their authority over the Bashu through the symbolism of milk giving (BHT 128, 132).

15. The prior investiture of those notables who invest the *mwami* may also reflect the danger involved in the accession ceremonies described below and the belief (reflected in other types of initiations and social transformations, such as those associated with circumcision and funeral rites) that only individuals who have gone through a similar process can safely come in contact with those undergoing the process for the first time. Thus only circumcised men can be present in the circumcision camp, and only people who have already lost a close relative can prepare the food and clothes and shave the heads of the family of a recently deceased man or woman. See Mary Douglas, *Natural Symbols* (1973), for a general discussion of this phenomenon.

16. For example, at funerals the wives and children of the deceased must put on clothes made of green banana leaves and eat cold food served on green banana leaves as part of the process by which they are cleansed and reincorporated into the homestead. Newly circumcised boys, who are in transition and thus in a ritually dangerous condition, live in huts that are constructed of green banana leaves. The umbilical cord of a newborn child is wrapped in a green banana leaf, and buried in 'cool,' moist soil at the foot of a banana tree in order to 'cool."

17. Beidelman (1966: 399) sees a similar process involved in the Swazi king's initial copulation with the ritual queen following his taking on the filth of the nation.

18. This association also occurs in proto-bantu: 'red' / *-kud-; 'to grow' / *-kuda. See C. C. Wrigley (1973: 219–34) for a discussion of the color red in Nyoro symbolic thought.

19. The prohibition against sex also applies to other activities that are susceptible to pollution. For example, if a person who has had sexual relations during the previous day enters the place where iron is being forged, the tool being forged will be weak. As noted above, the parallel between iron that is being forged and the *mwami* who is being invested is made explicit in the role of the *mwamihesi* during the accession ceremonies.

Sex is also prohibited during sacrificial rites. It is said that if the person performing the sacrifice has had sexual relations during the preceding day, the sacrifice will not be successful. Thus the sacrifice, like iron, and the *mwami* become susceptible to the ritual pollution from sexual activity.

20. Iron is potentially polluting because it comes from under the earth and thus from the world of the ancestors rather than the world of the living. The danger involved in moving between these two worlds or in moving objects from one to the other is reflected in the prohibition against digging deep holes except when burying the dead. This prohibition in turn clarifies the danger involved in the ritual burial of the *mwami* during the accession ceremony, as well as Bashu resistance to the digging of latrines during the colonial period. As a result of this danger, men who went into the ground to dig iron ore *(matale)* performed a special sacrifice to the spirit Muhima, "the protector." Despite this danger, iron tools can be used under normal circumstances if the proper land rituals are performed. However, they become a source of pollution when the land enters a state of ritual danger such as is brought on by the *mwami*'s investiture. For at such times the land, like the *mwami*, is more susceptible to pollution.

21. An invested *mwami* cannot wade across a river but must be carried. This is evidently because rivers are seen as collecting pollution, which is carried down the sides of ridges. Dangerous substances are always placed downhill from the homestead, and cleansing the homstead, as noted above, occurs in a downhill direction. A number of informants stated that the bouquet used in cleansing the ridge was thrown into a stream.

While the *mwami*'s presence is required at all major sacrifices for the land, he cannot himself perform the sacrifice because this action is ritually dangerous and because the sacrifical knife or spear, which is made of iron, is potentially polluting (see n. 20). He therefore employs a ritual surrogate called the *muheri*, who is generally the *mukulu*, to perform the actual killing of the sacrificial animal.

When the *mwami* goes on a trip, which in the past was a rare occurrence, he must be accompanied by a *mukumu*, who makes sure that his path is free from polluting substances and *balimu bavi*. Moreover, before the *mwami* can return to his royal compound, *kikali*, it must be cleansed of any dangerous substances that may have been placed there during his absence.

The *mwami* has to avoid women who are menstruating, murderers, and persons who have performed other ritually dangerous actions and have not yet been cleansed.

22. The significance of this coffin is unclear. While most informants likened it to a trough in which banana beer is fermented, known locally as an *muhe*, in the context of the *mwami*'s funeral, it is called a canoe, *bwatu*. Canoes are not presently part of the material culture of the mountain-dwelling Bashu. The idea of a *mwami* being placed in a canoe may therefore be a vestige of earlier mortuary practices that developed among the Bashu when they lived in the plains along the

lake shore and on the banks of the Semliki and its tributaries. Similar practices occur among the Hunde, Havu, and the Kiga of Kayonza and among a few southern Nande groups. It may therefore be a very old element of the region's political culture. The origins and dispersion of the practice are, however, impossible to determine with the data presently available. Richard Sigwalt (1975a) has begun to reconstruct the early political culture of the Western Lakes Plateau Region and this work may eventually allow us to gain a better understanding of the region's cultural history.

23. As one informant put it, "When the *mwami* is without his *mbita* the land is uncovered" (BHT 63).

24. See Packard (1980a) for a more extensive discussion of the cosmological position of women in Bashu society as it relates to the problem of misfortune.

25. I was told that in the past the wives of hunters could tame wild animals by standing naked on a hill overlooking the spot where the animal was located and crying out *Uli nyama y'omukali,* 'You are an animal of women!' This is said to have made the animal tame and easy to kill.

26. Following the accession ceremonies, the *mombo* was given a hill on which to live and over which she was chief. She was guarded by a number of young men, some of whom she might take as lovers. According to some informants, the *mombo* could not again meet the *mwami,* for she was regarded as a *mwami* herself and two *bami* cannot meet without disastrous consequences for themselves and for the land.

The *mombo's* role in the rebirth of *bwami* can perhaps be seen in terms of what Peter Ribgy (1968), describing the Gogo of Tanzania, calls "a re-reversal of time." According to Rigby, the Gogo see adverse ecological conditions as resulting from a reversal of ecological time, from 'time out of time.' To alleviate this situation, time must be "re-reversed." This is accomplished by a reversal of ritual roles, in which women perform ritual functions normally reserved for men. Following this suggestion, the Bashu *mombo* may contribute to the renewal of chiefship by re-reversing its temporal dimension, previously reversed by the death and burial of the *mwami.*

27. See chapter eight for a more detailed discussion of Bashu political development during the colonial era.

28. The sources for this table are as follows: Nyoro (J. W. Nyakatura, 1973; K. W., 1937; J. Roscoe, 1923; J. Beattie, 1972), Nkore (S. R. Karugire, 1972; J. Roscoe, 1923); Ganda (J. Roscoe, 1966; A. Kaggwa, 1934); Rwanda (M. d'Hertefelt and A. Coupez, 1964); Hunde (L. Viane, 1952b); Nyanga (D. Biebuyck, 1956, 1979); Havu (H. Verdonck, 1928); Shi (P. Colle, 1921; R. Sigwalt, 1975); Southern Nande (R. Debatty, 1951). Information on the Bashu and Bakonjo is based on my own field work.

29. This consistency of ideas and practices supports the conclusion reached by J.-P. Chretien (1977) (summarized in a review by Vansina [1978b]) that the sources of sacrality are found at the level of the whole society and its relations to nature. "The king then is owned by kingship but kingship is determined by the whole practice and understanding of ritual in society. Historians infer that kingship essentially is an outgrowth of older organizations, ideas, and rituals."

CHAPTER TWO: MIGRATION, SETTLEMENT, AND THE ROOTS OF BASHU CHIEFSHIP

1. According to Father L. Bergmans (personal communication), large sections of this forest have been cleared in the last twenty-five years. Muleke and

Kasongwere were still covered with forest when the Belgians arrived in 1923.

2. BHT 21, Kalembo Mbura, July 6, 1974; BHT 27, Rutava, June 6, 1974; BHT 31, Itaheri, June 24, 1974; BHT 44, Mughugha, July 10, 1974.

3. L. Bergmans (1970: 92–95). John Hart, a naturalist studying ecological relationships among the Bambuti Pygmies of the Ituri Forest, provided me with a great deal of information on the Bapere and Bapakombe, forest Bantu groups with whom the Bambuti interact and with whom he lived at several points during his research.

4. Edward Winter (1956: 6); M. Trowell and K. P. Wachsmann (1953: 6–7).

5. S. Lingier, "Étude sur la chefferie des Bambuba," Beni, 2–20–1932, typed manuscript located in the North Kivu file of the Ethnographic section of the Musée Royal de l'Afrique Centrale, Tervuren, Belgium.

6. Bashu migration traditions are in direct opposition to those of the Bakonjo. While the Bashu claim to have come from the east and in some cases the Ruwenzoris, Konjo traditions trace their origins to the west and often to Isale (see J. S. Matte, forthcoming, for a discussion of Konjo traditions). A strong indicator of early Sudanic settlement can be seen in the rainmaking techniques employed by the Bashu and their former neighbors in the Semliki Valley (see pages 67–68 for a description of these techniques).

7. BHT 110, Sindani Katsuva, Oct. 14, 1974; BHT 143, Musienene and Mughugha, Dec. 13, 1974.

8. Heinzelin (1957: 36) notes that while the Bantu-speaking population of the region around Ishango at the time of his excavation grew plantains, manioc, and beans, the grinding stones found at the level of earlier agricultural occupation at Ishango had been used for grinding millet. Other grindstones found by Charles M. Good (1972) were apparently used for grinding sim sim and may date from a later period (personal communication, Merritt Posnansky).

9. The rising power of Bahima groups in the upper Semliki Valley during the early years of the eighteenth century is indirectly attested to by Bashu traditions. The Bashu claim that the Bahima did not arrive in the Semliki Valley until the early eighteenth century, approximately ten generations ago. Yet traditions from Uganda and Rwanda indicate that Bahima groups had moved into Busongora as early as the thirteenth century (see Buchanan, 1974). Bashu traditions may therefore reflect the rising power of the Bahima rather than their initial settlement in the valley.

10. A. Willemart, "Rapport d'enquête sur la grande chefferie des Batangi," Lubero, 20 November 1929, AZL; F. Absil, "Rapport d'enquête sur la chefferie d l'Utwe (Bamate)," Lubero, 20 December 1927, AZL.

11. The secondary nature of this stimulus may explain why there are so few references to the disease in Bashu traditions of migration.

12. P. E. Joset, "Procès-verbal: réunion des notables Wasongora," Kasindi, 27–29 November 1936. Correspondence, AZB. John Ford notes that Zulu herdsmen in Natal avoided infection by moving their herds into the high veld during the dry winter months (Ford, 1979: 270).

13. BHT 27, Rutava, June 17, 1974; BHT 73, Kitsera, Kamavu, December 8, 1974; "Rapport de Reconnaissance," December 1911, Registre des Rapports Politiques, Kasindi, June 1906–June 1913, AZB.

14. Personal communication from John Hart. The validity of this scenario should be verified by the presence of archeological sites containing villages with parallel lines of rectangular huts similar to those of the Babira within the Bashu

region. To date, however, no archeological work has been carried out in the area.

15. Personal communication from John Hart.

16. See Brian Taylor (1962: 92) for joking relations among Bakonjo clans.

17. BHT 44, Maghugha, July 10, 1974; BHT 142, Kambere Kasuyire, December 12, 1974; a similar description based on the testimony of the Bashu rainmaker Kamali is found in Kyoswire Katembo (1974). See also Bergmans (1971: 90–91).
    Similar ideas about the origins of rain and the role of quartz crystals were recorded by Stuhlman (1894: 282) among the peoples of the upper Semliki Valley in 1891. More generally, Bashu ideas concerning the use of rainstones appear to be a variant, perhaps shaped by local geography, of ideas and practices that are widespread among the Central Sudanic–speaking peoples of northeast Zaire, northwest Uganda, and the southern Sudan (A. W. Southall, 1953: 376–79; P. T. Baxter and Audrey Butt, 1953: 118–19, 124; C. G. Seligman, 1932: 131–32, 280–89). Bashu ideas are particularly close to those of the Madi peoples of northwest Uganda, where a similar identity exists between the digging-stick weights, the earth, and female gender, on one hand, and between crystals, solidified rain, and male gender, on the other. Moreover, as in the Bashu region, Madi rainstones are smeared with fat and left in the sun before being immersed in water. The similarity between Bashu rainmagic and that of these Central Sudanic groups, combined with the apparent absence of similar ideas elsewhere among the lacustrine Bantu groups of the Western Lakes Plateau Region, suggests that Bashu rainmagic may be of Central Sudanic origin. The absence of evidence of recent contact between the Bashu and these Central Sudanic speakers suggests that these ideas are of considerable antiquity, perhaps having been introduced into the upper Semliki region during the eighth or ninth century, when, according to Buchanan (1974: 44–63), Madi-related groups moved into the region.

18. These alliances are discussed in chapter four.

CHAPTER THREE:

THE POLITICAL TRANSFORMATION
OF THE UPPER SEMLIKI VALLEY REGION

1. For a discussion of these events, see S. Karugire (1972).

2. This reconstruction of events is based primarily on Zaire sources (BHT 165, Kalongo Mohambo, February 2, 1975); Reconnaissance Report, October 1911, Renseignements Politiques-Kasindi, 1908–1913, p. 53, AZB; "Rapport d'enquête de la chefferie Bahema," Registre des Renseignements Politiques-Beni, 1928–1937, p. 334, AZB. Kamuhangire (1972: 11) tentatively dates the founding of Bugaya between 1715 and 1742 on the basis of genealogical information on Bugaya collected in Uganda. While Kamuhangire's chief list for Bugaya largely corresponds to those provided by Zaire sources, the dates he gives to each ruler appear to be off by some seventy-five years. The last ruler on his list, Kalongo s/o Masidongo, whom Kamuhangire dates to 1823–50, was still alive in 1928. Adjusting his list to this date would place the foundation of Bugaya between 1790 and 1823, which corresponds to Zaire sources.

3. E. C. Lanning (1954: 24–30). Personal communication, E. R. Kamuhangire, December 31, 1974. I am grateful to Dr. Kamuhangire for information concerning the history of the Babito rulers of Busongora.

4. BHT 53, Sabuni Kasango, July 24, 1974; BHT 108, Mohorodamo Muhindo, October 14, 1974; BHT 165, Kalongo Mohambo, February 2, 1975.

5. The above discussion of Bakingwe activities in the Katwe region is based largely upon E. R. Kamuhangire's important study of the economic and social history of the Southwestern Uganda Salt Lakes Region (1975: 74–79).

6. BHT 110, Sindani Katsuva and Musinene, October 22, 1974.

7. BHT 143, Mughugha and Musinene, December 13, 1974; BHT 128, Siyindwe, November 19, 1974; BHT 78, Mbugha Kahindo, August 8, 1974; BHT 79, Saa II, August 15, 1974.

8. BHT 105, Meso, November 10, 1974 (Lisasa); BHT 76, Mukumbwa Kahindo, August 13, 1974 (Luvere); numerous traditions describe the arrival of Kavango's family, whose descendants presently rule the Bashu chiefdoms.

9. See Jan Vansina (1973: 220–40) for a discussion of this problem with reference to the Kuba of Zaire.

10. There are a small number of Bashu-related peoples under Bamoli leadership living at the fisheries at Kyavanyonge, on the north shore of the lake. The ancestor of the present Bamoli chief of the settlement settled in the Semliki Valley during the early years of the nineteenth century. There are also some Basongora peoples living along the edge of the Ruwenzori Mountains near Kasindi. Bakonjo/Nande settlements on the western slopes of the Ruwenzoris have in recent years begun to spread over the east bank of the Semliki, whereas the west bank has remained uninhabited.

11. Registre des Renseignements Politiques-Semliki, 1924–27, p. 44, AZB; Rapport d'enquête de la chefferie Bahema, Registre des Renseignements Politiques-Territoire de Beni, 1928–37, p. 334, AZB.

12. BHT 112, Kahembe, October 28, 1974.

13. H. Daniels to M. Henne, Beni, August 24, 1927, Correspondence, AZB.

14. Reconnaissance, June 1913, Renseignements Politiques-Kasindi, p. 61, AZB.

15. BHT 53, Sabuni Kasango, July 24, 1974; BHT 56, Katsuva, July 26, 1974; BHT 76, Mukumbwa Kahindo, August 11, 1974; BHT 165, Kalongo Mohambo, February 2, 1975.

16. A major gap exists in our knowledge of Kisaka. While I was able to collect information from Basongora living in Zaire, I was unable to enter Uganda. Kamuhangire's research on the social and economic history of the Salt Lakes Region did not include extensive questioning about political culture and royal ritual. There is some evidence that points to Bakingwe influence on Bashu political culture. This is based on a comparison of certain Bashu practices with those of the Kigezi kingdom of Kayonza, which, like the Bakingwe political system in Busongora, was evidently led by a section of the Barenge clan. First, the place where a *mwami* is invested is called *Isingiro* in Kayonza and among the Bashu and other Nande groups. This term does not occur in the ethnographic records of other lacustrine Zaire states, nor in western Uganda. Similarly, the use of the term *bakaka* to refer to royal buriers of a deceased *mwami* is apparently limited to Busongora, Kayonza, Banisanza, and the Nande. In addition, certain practices associated with the *bakaka*'s role are similar in all these areas. Included among these are the public performance of sexual relations between the *bakaka* and specially designated ritual wives after the burial duties have been completed. A second practice involves the *bakaka*'s retreat to a specially designated hill, where the *bakaka* will live from then on, and the future avoidance of the *bakaka* by the new *mwami*. The distribution of these elements combined with Barenge movements in the Western Rift Valley suggests that the presence of these elements among the Bashu may be traced to Barenge influence on Babito political culture

in Busongora and its subsequent importation to the Mitumbas by the Babito and Bamoli (see M. R. Rwankwenda, 1972: 124–33, for material on Kayonza royal ritual). Finally, two other elements of Bashu political culture may be traceable to Barenge influence. The first is the occasional use of the term *banyigbinya* to describe members of the royal family. Elsewhere, in Rwanda, Ndorwa, Bugesera, and Nkore, this term has similar usage (see J. Vansina, 1962: 43). The second element is the royal practice of keeping baboons as domestic animals. This practice is followed by the kings of Rwanda. Neither practice occurs among the southern Nande. It is, of course, possible that these various elements were introduced into the region by pastoralists who moved into the Busongora and the upper Semliki from Nkore, Ndorwa, and Rwanda. Again, more data on the political culture of Kisaka would help clarify this problem.

17. BHT 165, Kalongo Muhambo, February 12, 1975.

18. Rapport d'enquête sur la chefferie Bahema, Registre des Renseignements Politique-Beni, 1928–1937, p. 334; Reconnaissance, October 1911, Registre des Renseignements Politiques-Kasindi, 1908–1913, p. 52, AZB; F. Lugard (1959: 273).

19. BHT 15, Mushyakulu, April 29, 1974.

20. BHT 69, Katafali, August 9, 1974; Reconnaissance, October 1911, p. 53.

21. In Bunyoro, succession struggles were regular events during the nineteenth century. Mukama Kyavambe III (1786–1835) came to power by killing his brother. Olimi V Rwakabale ruled only five years before Kamurasi rebelled and killed him in 1852. Kamurasi's eventual successor, Kabarega, fought from 1869 to 1871 before defeating his rivals for the throne (see G. N. Uzoigwe, 1970). Similarly, from the death of Kaboyo around 1850 until the British installation of Kasagama in 1891, a long series of succession wars occurred in Toro, stimulated in part by outside intervention from Buganda, Bunyoro, and Nkore. During this period eight different Bakama sat on the throne of Toro. The princes of Kisaka were frequently involved in these struggles, either siding with one or another rival for the throne or attempting to take advantage of the conflicts to establish their own independence (see Sir George Kamurasi Rukidi, "The Kings of Toro," pp. 12–17).

22. The following is a list of the major raids against Busongora during this period:

Buganda raid during the reign of Kabaka Suna II, ca. 1825–52
Toro raids   "   "   "   "   Mukama Kaboyo, ca. 1830s–1840s
Nkore raid   "   "   "   "   Mugabe Mutumbuka, ca. 1852–78
Buganda raid   "   "   "   "   Kabaka Mutesa, ca. 1871
Bunyoro raid   "   "   "   "   Mukama Kabarega, ca.1880s and early 1890s
Toro raid   "   "   "   "   Mukama Kasagama, ca. 1894

23. See chapter six for a discussion of these events.

24. Rapport d'enquête sur la chefferie Bahema, p. 324. It is difficult to evaluate the impact of rinderpest on the cattle population of the upper Semliki Valley. There is no mention of the disease in Bashu traditions or in the few traditions collected from former herdsmen. Scott Eliot, who visited the valley in 1894, noted that cattle were thriving in the region. Belgian records beginning in 1897 are also silent about the disease. John Ford (1971: 173) suggests that either the disease was late in arriving or the cattle there made a rapid recovery. It would

appear from these references, or absence thereof, that if the disease did hit the region, its effects were limited. One reason for this may have been the political instability and warfare that dominated the region at the end of the nineteenth century. The threat of loss of cattle through raiding forced the pastoralists of the region to hide their cattle in small herds along the edges of the mountains. Dispersing the cattle in this fashion may have lessened the impact of the disease.

25. See chapter four for a discussion of these alliances.

26. The disparity between the livestock wealth of the plains dwellers and that of the mountain folk is attested to by Bashu traditions and by the fact that the Semliki Valley served as a reservoir from which the Bashu built up and replenished their herds. When drought or disease killed off livestock in the mountains, the mountain people acquired new animals from the plains. This disparity in livestock wealth is also indicated by the statement of a Muhima informant that the Bahima used to call the people of the mountains *avai-ira* ('people without wealth'), "because they were poor while we in the plains had many cattle and goats" (BHT 112, Kahembe, October 28, 1974).

27. BHT 70, Kahindo Mbita, August 10, 1974; Reconnaissance, October 1911, p. 53.

CHAPTER FOUR: RITUAL CHIEFSHIP
AND THE POLITICS OF DOMINATION IN ISALE

1. Some data are available on the status of early allies. Berger (1980) indicates the importance of *mbandwa* mediums as allies of the Babito in Bunyoro. See also A. Roberts (1976), S. Feierman (1974), D. Cohen (1977).

2. BHT 44, Mughugha, July 10, 1974; BHT 90, Mughugha, September 14, 1974; BHT 78, Saa Mbili, August 15, 1974; BHT 77, Mate Kiniki, August 14, 1974; BHT 145, Katembo Maniere, December 17, 1974; BHT 133, Kasuki, November 26, 1974; BHT 57, Kasango, July 27, 1974; BHT 58, Kambale Kahingani, July 29, 1974; BHT 163, Kasogho Kambale, February 11, 1975.

3. All the traditions that describe the coming of the Babito begin with references to their pastoral origins and with the fact that they possessed many cattle. Examples of this can be seen in variants of the Muhiyi traditions presented in the introduction.

4. BHT 44, Mughugha, July 10, 1974.

5. BHT 165, Kalongo Muhambo, February 12, 1975; BHT 108, Mohorodamo, October 14, 1974. This claim corresponds with that made by the Bahema groups who expanded into the territory of the Lendu in the lower Semliki Valley. According to Lobho-lwa Djugudgugu (1974: 34), the Bahema claim to have acquired their rainmaking skills from the Lendu.

6. BHT 106, Sivindere Virere, October 13, 1974; BHT 24, Kassemengo, June 12, 1974. Kavango's descendants may also have used rain pots, for the pot *(riregba ly'ovusyano)* in which the *mwami* keeps his ritual seeds *(mikene)* is also called *riregba ly'esyombula* in one tradition, *syombula* being the plural form of *mbula,* 'rain.' Since early Babito chiefs did not "own" the *mikene* but acquired them from 'healers of the land,' it is possible that the *riregba ly'ovusyano* represents the joining of the use of rain pots with the earlier practice of distributing *mikene.*

7. BHT 165, Kalongo Muhambo, February 2, 1975.

8. The mechanics of the salt trade are described in a number of Bashu traditions. See especially: BHT 17, Kavota Mugasi, June 1, 1974; BHT 22, Kasereka Masikini, June 7, 1974; BHT 53, Sabuni, July 1, 1974.

9. Fifty fiber rings made up a measure called a *kine*. Two *kine*, or 100 rings, equalled a *mughanda*. I was given the following exchange equivalents for *vutegha* in Isale: 1 *mughana* = 1 chicken, 10 *mighanda* = 1 goat, 2 *mighanda* = 1 hoe blade, 2 *mighanda* = 1 goat skin, 1 *mughanda* = 1 basket of beans, 1 *kine* = 1 *luhinda* (iron ring worn on the ankle), 2 *ekine* = 1 *riregha* (large pot), 1 *kine* = 1 *kitere* (large basket). Presumably these equivalents varied through time and from place to place, and not all informants gave the same equivalents. It is clear, however, that *vutegha* did serve as a form of local currency.

10. BHT 17, Kavota Mugasi, June 1, 1974; BHT 53, Sabuni, July 1, 1974. One informant suggested that Bashu access to the salt lake resulted from ritual assistance that one of Kavango's descendants, Vyogho, gave to the chiefs of Busongora at the end of the nineteenth century. Other informants, however, insist that they have always had this privilege.

11. BHT 159, Kasondiva, February 1, 1974; BHT 141, Katsuva, December 12, 1974; BHT 52, Mbayakulya Kamabo, Mwendokolero, and Vahekeya, July 22, 1974. According to A. Moeller (1936: 42), Kavango's ancestor Maherere married a Mutangi woman in Busongora before the family moved to the west bank of the Semliki.

12. BHT 106, Sivindere Virere, October 13, 1974.

13. BHT 52, Mbayakulya et al., July 22, 1974.

14. BHT 88, Kalendero, September 12, 1974.

15. Some informants claim that Kavango also married a Muhera woman from Maseki, who bore a son named Kahese. There is, however, strong evidence that suggests that Kahese was the son not of Kavango but of his grandson Luvango. See note 30 for a fuller discussion of this problem.

16. BHT 54, Valymugheni, July 25, 1974. Original forest dwellers evidently played a similar role in the accession ceremonies among the Nyanga, Hunde, and Bashi and in Rwanda.

17. BHT 64, Katsoperwa, August 8, 1974. Some traditions claim that the biological mother of Mukunyu's successor, Visalu, was a Mutangi. This is apparently because Mukunyu's *mombo* was a Mutangi. Thus sociologically, Visalu's mother was a Mutangi. There is in general a tendency for the clan of the biological mother to be forgotten over time, though it is always remembered by the clan itself.

18. See chapter three, p. 77, for a discussion of the role of Itsinga in Bashu accession ceremonies.

19. BHT 63, Mutunzi, August 5, 1974; BHT 65, Matsinda, August 6, 1974; BHT 68, Vahemba Kahese, August 8, 1974. F. Van Rompaey, "Compte rendu de la conversation tenue le 14 Août 1945," AZB.

20. This conclusion is based on Bashu traditions and on the claims made by related Babito chiefs on the western slopes of the Ruwenzoris who have not adopted the *mombo* institution and continue to follow the practice of primogeniture. BHT 70, Kahindo Mbita, August 13, 1974; BHT 106, Sivindere Virere, October 13, 1974; BHT 107, Yongele, October 13, 1974.

21. BHT 166, Mwendokolero, February 14, 1975 (guardian of Mukunyu's tomb).

22. While eastern Malio presently resembles Isale in terms of its physical environment, there is strong evidence that it was more heavily forested at the time of Kamesi's settlement in the region. First, the name of the place where Kamesi was raised, Musitu, means 'forest' in Kinande. Second, several families of 'clearers of the forest' located in eastern Malio and northern Ngulo claim that their

ancestors cleared the forest in the region between six and seven generations ago, and thus around the time of Kamesi's settlement in the area. In addition, traditions from the people of Mwenye, located in Isale near the present border with Malio and northern Ngulo, refer to raids on their settlements by Basumba from Malio during Luvango's reign (see below).

23. BHT 44, Mughugha and Katembo Maniere (Muswaga of Katikale), July 10, 1974; BHT 158, Kalani Masisa, January 30, 1974 (guardian of Luvango's tomb).

24. BHT 166, Mwendokolero, February 14, 1975; BHT 27, Rutava Muletya, June 17, 1974; L. Bergmans (1974: 23).

25. BHT 50, Kasayi, July 18, 1974; BHT 159, Kaloni Masisa, January 30, 1974.

26. BHT 35, Balikwisa, June 27, 1974; BHT 41, Kasereka Zamani, July 5, 1974.

27. BHT 158, Masisa Kalani, January 30, 1975; BHT 135, Rutava, November 29, 1974.

28. There are conflicting traditions concerning Kivoto's parentage. While the Bahera of Ngukwe and Kivoto's descendants claim that Kivoto's mother was a Muhera (BHT 74, Kinyandali, August 13, 1974; BHT 24, Kassemengo, June 12, 1974), the descendants of Kivoto's brother Muhoyo claim that Kivoto's mother came to Luvango's home looking for food during a famine and that she was caught stealing food and locked in a hut. While a prisoner in the hut she became pregnant, which, given her seclusion at the time, caused the people to be amazed and to wonder who the father was. When the woman gave birth, the child resembled Luvango, who therefore adopted the child as his own and eventually chose him to be his successor. A second tradition states that Kivoto was captured by Luvango during his war with the Banisanza of Lisasa (see below). While it is possible that Kivoto was adopted by Luvango, these traditions may simply be attempts to discredit Kivoto and question his right to succession. As will be seen in chapter five, Kivoto was involved in a succession struggle with his brother Muhoyo following Luvango's death.

29. Kahese's descendants claim that Kahese was the son of Kavango rather than Luvango. This conflicts with an early tradition collected by the Belgians, in which Kahese is said to have been Luvango's son (J. A. Offenheim, "Note historique sur le Maseki et compris le groupe Kakuse," June 25, 1935, AZB). In addition, the genealogy for Kahese's descendants contains only two names that are repeated: Kahese I f/o Rukanda I f/o Kahese II f/o Rukanda II f/o Kahese III, who is still alive. This in itself is not cause for suspicion, since a *mwami* is thought to inherit the power of his grandfather, and grandsons often bear the names of their grandfathers. However, I was unable to discover burial sites for Rukanda I or Kahese II. Moreover, nothing is remembered about their lives. There is reason to believe, therefore, that these two names have been inserted into Kahese's genealogy to give it additional antiquity. This interpretation is consistent with the activities of Kavango and Luvango. From what we know about Kavango, it seems highly unlikely that he would have established an alliance with this distant mountain group. On the other hand, such an alliance is consistent with Luvango's widespread involvement in mountain politics.

30. BHT 97, Kavulo Masikini, October 2, 1974; BHT 98, Kavwaro, October 4, 1974; BHT 167, Kavwaro, February 15, 1975.

31. BHT 163, Kasogho Kambale, February 11, 1975; BHT 44, Mughugha, July 10, 1974; BHT 27, Rutava, June 17, 1974.

32. BHT 27, Rutava, July 17, 1974; BHT 158, Masisa Kalani, January 30, 1974.

33. BHT 90, Muhugha, September 14, 1974; BHT 164, Kambale Vinyanzi Vighala, February 11, 1975; Van Rompaey, "Compte rendu ... le 14 Août 1945," AZB, p. 1.

34. It could, of course, be argued that the alliance was equally beneficial to Mbopi, for it improved his position vis-à-vis Kisere while increasing the division between Luvango and Visalu. In the end, however, the alliance was more beneficial to Luvango than to Mbopi, for Luvango's authority, unlike that of Mbopi, was based more on his own success at alliance building than on a claim to his father's authority. Thus Luvango's dispute with Visalu was less debilitating for him than Mbopi's dispute with Kisere was for either of them.

35. BHT 35, Valikwisa and Vatembwa, June 27, 1974; BHT 93, Kisoro, September 21, 1974; BHT 163, Kasogho Kambale, February 11, 1974; Van Rompaey ("Les Régles coutumières régissant la succession des chefs Bashu," February 10, 1945, p. 6, AZB) suggests that Luvango was content to remain as *mukulu* until after Visalu's death and that he then usurped the position of Visalu's son, who fled to the Banisanza. While this is possible, my own data indicate that the succession struggle occurred during Visalu's lifetime. It is conceivable, of course, that the truth lies somewhere in between—that Luvango killed Visalu and that Molero then fled to Lisasa.

36. BHT 59, Paul Muhindo, August 3, 1974; BHT 75, Mukumbwa Kahindo, August 13, 1974; BHT 140, Kyabale, December 11, 1974.

37. Van Rompaey, "Compte rendu ... August 14, 1945," p. 1, AZB.

38. BHT 158, Masisa Kalani, January 30, 1974. The Batangi of Musindi also invested the *mwami* of the Bashu chiefdom of Vuhimba in 1960. Prior to this the Bashu of Vuhimba had been subjects of the Bamate of Utwe and had not invested a *mwami* since the time of Kihimba, in the middle of the nineteenth century. In order to acquire knowledge of the proper investiture practices, representatives from Vuhimba went to Kisuki, a descendant of Luvango, in Isale. Kisuki told them to go to the Batangi of Musindi, for it was they who had taught his ancestors the proper procedures. (BHT 129, Kateme, November 19, 1974; BHT 130, Muhindo Kambiro, November 20, 1974; BHT 132, Vutsumba, November 20, 1974.)

39. See Tables 2 and 3 for distribution data.

40. BHT 13, Mushyakulu, April 11, 1974; BHT 125, Kawere Kahita, November 16, 1974.

41. See J. Claessens (1929: 4–56) for an early description of Bashu cultivation.

42. According to several informants, the mother of Luvango's chosen successor, Kivoto, came to Luvango's home during a period of famine seeking food. See note 28, above. Other traditions from two separate areas refer to the fact that the people of these areas were forced to take sorghum to Luvango as tribute. Sorghum is grown by the Bashu primarily as a hedge against adverse climatic conditions because it is more resistant to such conditions. These traditions therefore attest to the existence of famine conditions during Luvango's reign.

CHAPTER FIVE: SUCCESSION, FAMINE, AND POLITICAL COMPETITION IN ISALE

1. Jan Vansina (1962), using a different classification, refers to these political systems as "regal aristocracies."

2. This system of succession obviously contrasts with that found in a number of other African societies in which political authority is tied to the problem of ecological control, and interregna are consequently kept as short as possible or eliminated, either symbolically, by keeping the king's death secret or by replacing the king with a temporary surrogate (as occurs among the Bunyoro and Shilluk), or by coordinating the burial of the old chief to occur simultaneously with the investiture of the new ruler (as occurs in Shambaa). The presence of extended interregna among the Bashu is, I suggest, a reflection of the specific nature of Bashu political ideas.

3. BHT 98, Kavwaro and Kavulo Masikini, October 4, 1974; BHT 73, Kitsera Kamavu. August 12, 1974.

4. BHT 35, Balikwisa, June 27, 1974. Muhoyo reinforced his alliance with the Baswaga of Ngulo by taking his *mombo* from this group.

5. BHT 35, Balikwisa, June 27, 1974; BHT 123, Rutava, November 13, 1974.

6. BHT 90, Mughugha, September 14, 1974.

7. BHT 163, Kasogho Kambale, February 11, 1974; BHT 53, Kayangani, July 29, 1974.

8. BHT 20, Kasongya, June 6, 1974; BHT 87, Kasongya, September 3, 1974. Muhoyo's family sends a goat to Muhiyi's family prior to their accession. This goat is called *mbene y'embikulu* and announces the intention of Muhoyo's family to crown a chief. This goat was first given by Muhoyo to Muhiyi's grandson Nunyu-Iremba.

9. There are numerous references to famines in Bashu traditions that follow a set formula: a chief was invested and performed the appropriate sacrifices for the land and saved the people from famine. These references lack detail and may well be cultural metaphors for periods of political change.

10. BHT 43, Kitanda Viavo, July 6, 1974.

11. BHT 43, Kitanda Viavo, July 7, 1974; BHT 40, Kambere Vunghove, July 3, 1974.

12. BHT 41, Zamani Kasereka, July 5, 1974.

13. BHT 123, Rutava, November 13, 1974; BHT 163, Kasogho Kambale, February 11, 1975.

14. BHT 168, Mukanirwa, February 17, 1975; BHT 62, Kadyadya, August 5, 1974. In Toro, Basangwa refers to "the people who were found" and thus to the autochthonous status of certain groups (Richards, 1960: 130). While the 's' to 'sy' sound shift is regular from Lutoro to Kinande, I have not found this meaning attached to the name Basyangwa among the Bashu.

15. BHT 46, Kavwaro, July 13, 1974; BHT 51, Metya Kataka, July 19, 1974; BHT 53, Mutama Sabuni, July 24, 1974; BHT 5, Mombokani Wasokundi, March 24, 1974.

16. According to Mutsora's descendants, Mutsora moved to Vwambala in the plains following Luvango's death in order to lay claim to that portion of his father's herds to which, as firstborn son, he was entitled. He subsequently settled in the lower valley of the Talia, where he was welcomed by a section of the Bakira clan. BHT 21, Kalemo Mbura, June 7, 1974; BHT 26, Kanigha Kambale et al., June 15, 1974.

17. BHT 19, Mombokani Wasokundi, June 5, 1974; BHT 46, Kavwaro, July 13, 1974; BHT 40, Vunghove, July 3, 1974; BHT 50, Kasayi, July 18, 1974.

18. The inability of Mutsora's descendants to remember the details of this

alliance may reflect more recent political relations in the region. See below for a discussion of these relations and their impact on the region's traditions.

19. BHT 50, Kasayi, July 18, 1974.

20. BHT 19, Mombokani, June 5, 1974.

21. BHT 19, Mombokani, June 5, 1974.

22. BHT 50, Kasayi, July 18, 1974.

23. BHT 19, Mombokani, June 5, 1974.

24. BHT 50, Kasayi, July 18, 1974.

25. BHT 51, Metya Kataka, July 19, 1974; BHT 53, Mutamo, July 24, 1974; BHT 46, Kavwaro, July 13, 1974; Réunions du conseil des chefs Bashu, 18 November 1943, AZB.

26. BHT 46, Kavwaro, July 13, 1974.

27. For a discussion of beer drinking as a cultural metaphor in Bashu traditions, see chapter seven, p. 149 and n. 27.

28. BHT 137, Kahindo Musumba, December 12, 1974.

29. BHT 40, Kambere Vunghove, July 3, 1974; BHT 50, Kasayi, July 18, 1974.

30. BHT 46, Kavwaro, July 13, 1974; BHT 43, Kitanda, July 6, 1974; BHT 50, Kasayi, July 18, 1974.

31. BHT 156, Longo, January 27, 1975; BHT 167, Kavwaro, February 15, 1974.

32. BHT 42, Mombokani, July 6, 1974; BHT 19, Mombokani, June 5, 1974.

33. BHT 64, Katsoperwa, August 6, 1974; BHT 52, Mbayakulya Kamabo, July 22, 1974.

34. BHT 159, Kasodiva, February 1, 1975. For a discussion of early Babito-Batangi associations see chapter four, pp. 90–91.

35. BHT 88, Kalendero Kisaha, September 12, 1974; F. Van Rompaey, "Compte rendu de la conversation tenue le 14 Août, 1945," p. 2, AZB.

36. BHT 54, Valymugheni, July 25, 1974.

27. BHT 164, Kambale Vinyanzi Vighala, February 11, 1975; F. Van Rompaey, "Compte rendu ... 14 Août, 1945," p. 1, AZB.

38. BHT 64, Katsoperwa, August 6, 1974.

39. Ibid.

CHAPTER SIX: TRADE, FIREARMS, AND THE POLITICS OF ARMED CONFRONTATION IN THE UPPER SEMLIKI VALLEY REGION

1. S. Feierman (1974: 145–68). Feierman notes elsewhere (1972: 304) that Semboja's son, while adopting certain elements of western culture, "was careful to have the houses of Vugha built according to the pattern of earlier kings. The unchanging town of Vugha was too important a symbol of sovereignty among the Shambaa to be altered. The King made no attempt to bring large groups of traders to Vugha, and commerce was concentrated at Semboja's Mazinde."

2. At the same time the price of ivory rose steadily during the nineteenth century, making it economical to transport ivory over great distances (Curtin et al., 1977: 393).

3. For activities of Banyoro armies among Bashu, see BHT 18, 24, 26, 35, 44, 52, 60, 66, 78, 106, 112.

4. BHT 26, Kangigha Kambale et al., June 15, 1974; BHT 58, Kayighani Kambale, July 29. 1974; BHT 92, Kakule Yoani, September 19, 1974.

5. BHT 58, Kayighani Kambale, July 29, 1974; BHT 28, Kanigha Kambale, June 15, 1974; BHT 92, Kakule Yoani, September 19, 1974.

6. BHT 165, Kalongo Mahombo, February 2, 1975.

7. Karakwenzi had already arrived at Katwe when Stanley visited there in 1889. Stanley (1890: 334). I wish to thank Dr. Freedman for providing this information on Karakwenzi.

8. BHT 59, Paul Muhindo, August 3, 1974. The use of armed force combined with proselytization for Nyavingi provided a strategy for political expansion in Kigezi at about this same time and led to the diffusion of the Nyavingi cult in Kigezi (F. Geraud, 1972: 23). Karakwenzi's use of Nyavingi as an instrument of legitimation, combined with an alliance with the people of Busongora in opposition to Babito domination, suggests that Nyavingi was initially a focus of populist opposition to aristocratic pastoral rule in this region, as it was elsewhere outside Ndorwa. In the end, however, it served to legitimize a new aristocracy. See Freedman (1979) for a discussion of the wider history of the Nyavingi cult in the Western Lakes Plateau Region.

9. It was in 1894 that the newly installed king of Toro, Kasagama, launched a major military expedition into Busongora and the upper Semliki Valley to enforce his authority in the region and acquire cattle. G. F. Scott Eliot (1896: 105).

10. Joset, "Extraits de la rapport du Commissaire de District Hackers. Organisation de la population du Territoire de la Semliki," 1927, De Ryck Collection (46–1).

11. For Kirongotsi's activities in northern Isale, see BHT 35, Balikwisa, June 27, 1974; BHT 52, Mbayakulya Kamabo, July 22, 1974; BHT 57, Kasongo, July 24, 1974.

12. BHT 36, Mbalema Matavugha, June 29, 1974.

13. BHT 56, Katsuva, July 26, 1974; BHT 92, Kakule Yoani, September 19, 1974; BHT 105, Meso, October 11, 1974.

14. BHT 58, Kayighani Kambale, July 29, 1974; BHT 44, Mughugha, July 10, 1974.

15. Bergmans (1970: 37–38); BHT 136, Musubao Kighogholo, December 3, 1974.

16. BHT 35, Balikwisa, June 27, 1974; BHT 103, Akili Tsongo, October 10, 1974; BHT 104, Nzai, October 11, 1974; Bergmans (1970: 47).

17. Joset (1939: 11); "Papers Relating to the Execution of Mr. Stokes in the Congo State," British Parliamentary Papers, 1896, LIX. The original post of Beni, located in the valley, was subsequently moved into the bordering highlands, following the outbreak of sleeping sickness in 1905.

CHAPTER SEVEN: "NEW MEN"
VERSUS RITUAL CHIEFS IN ISALE

1. Kivere's descendants deny that such a dispute ever occurred and, in fact, claim that Kivere was Molero's son. They further claim that the struggle that did occur at this time was not over the question of succession but revolved around

the fact that Kivere had seduced one of Molero's wives. This claim, however, is denied by Vukendo's descendants and other informants and is inconsistent with other genealogical data. BHT 52, Mbayakulya et al., July 22, 1974; BHT 54, July 25, 1974; BHT 48, Kule Mikongo, July 14, 1974.

2. Kivere's descendants explain the loss of Batangi support by claiming that the Batangi grew tired of giving their women to Visalu's descendants. BHT 52, Mbayakulya et al., July 26, 1974.

3. BHT 30, Katamu and Paluku, June 22, 1974; BHT 50, Kasayi, July 18, 1974.

4. The verb *-tambika* also carries the meaning 'to take in one's arms like a baby' (Pauline Fraas, 1961: 106). The Batangi of Lulinda describe the process as follows: "At the investiture of the *mwami* we must find him a wife. Even if the woman is not from Lulinda she must pass through Lulinda on her way to the *mwami* and sit on our knees." BHT 159, Kasodiva, January 2, 1975; BHT 167, Kavwaro, February 15, 1975; BHT 43, Kitanda Viavo, July 6, 1974.

5. BHT 43, Kitanda, July 6, 1974; BHT 26, Kangighi, June 15, 1974; BHT 34, Sirawayo, June 26, 1974.

6. BHT 133, Kisuki, November 26, 1974; BHT 168, Mukanirwa, February 17, 1975.

7. BHT 53, Mutamo and Sabuni, July 24, 1974; BHT 56, Katsuva, July 26, 1974.

8. Correspondence from F. van Rompaey, Kalenghea, 27 October 1944, 80/A.I.M.O./B, AZB; BHT 45, Kaghoma Muhashu, July 13, 1974.

9. BHT 24, Kassemengo, June 12, 1974; BHT 42, Mombokani, July 6, 1974. See I. Karp (1980) for a fascinating discussion of beer drinking as an expression of commensality among the Iteso of Kenya. See also note 27, below.

10. BHT 32, Mombokani, June 24, 1974; BHT 48, Kule Mikongo, July 16, 1974. For information on the market at Muvulia: BHT 40, Kambere Vunghove, July 3, 1974; BHT 18, Kasongya Mitsumba, June 4, 1974; Kavota Mughasi, June 1, 1974.

11. BHT 32, Mombokani, June 24, 1974; BHT 93, Kasoro, September 21, 1974.

12. There is some evidence that Tsombira may have also raided Isale on his own. BHT 44, Mughugha, July 10, 1974; BHT 56, Katsuva, July 26, 1974; BHT 32, Mombokani, June 24, 1974.

13. The first date is based on the fact that Kasumbakali could not have acquired firearms from Tsombira before 1892. Tsombira himself acquired them from Karakwenzi, who only had a few firearms when Lugard arrived in 1891 and acquired most of his guns from Stokes sometime between 1891, when Stokes entered the region, and 1895, when Stokes was captured by the Belgians. The closing date is based on traditions that attribute Kasumbakali's decline to the onset of drought and a plague of jiggers in 1897. (Parliamentary Papers Relating to the Execution of Mr. Stokes in the Congo State, 1896, LIX. See especially F. Lugard, Memorandum on the case of C. Stokes, November 27, 1895.)

14. BHT 93, Kasoro, September 21, 1974; BHT 32, Mombokani, June 24, 1974.

15. BHT 51, Mutamo and Sabuni, July 24, 1974; BHT 59, Paul Muhindo, March 8, 1974; BHT 88, Kalendero Kisahi, September 12, 1974.

16. BHT 50, Kasayi, July 18, 1974; BHT 74, Kinyandali, August 13, 1974.

17. BHT 50, Kasayi, July 18, 1974; BHT 32, Mombokani, June 26, 1974.

18. BHT 26, Kangighi, June 15, 1974.

19. According to Bergmans (1974: 29), Kasumbakali also hanged victims. BHT 78, Kahindo Mbugha, August 15, 1974; BHT 36, Mbelema Tavugha, June 29, 1974.

20. BHT 53, Mutamo and Sabuni, July 24, 1974; BHT 56, Katsuva, July 26, 1974. According to Katsuva, Kasumbakali cut off Maha's hands. This evidently was the cause of his death.

21. BHT 47, Mokamoya, July 15, 1974; BHT 41, Zamani Kasereka, July 5, 1974.

22. L. Bergmans (1974: 29); BHT 78, Kahindo Mbugha, August 15, 1974; BHT 69, Katafali et al., August 9, 1974; BHT 59, Paul Muhindo, August 3, 1974; BHT 77, Mate Kiniki, August 14, 1974.

23. This date is based on several pieces of evidence. First, the onset of drought is said to have caused Karakwenzi to abandon his settlement in Maseki and return to the plains, where he attacked the Belgian station at Kasindi soon afterwards. This attack occurred in June of 1897. Second, the drought was evidently widespread, affecting all of Bunande as well as Busongora. Father Geraud records a major famine in Kigezi in 1897, and it is likely that this was caused by the same drought that affected neighboring regions. Finally, at least one tradition associates the drought with the passage of the Batetela mutineers, which occurred in 1897. BHT 163, Kasogho Kambale, February 11, 1975.

24. See, for example, BHT 31, Itaheri, June 24, 1974; BHT 69, Katafali Muhindo and Mwaholu, August 9, 1974. A similar view of the recent origins of goats also exists among the Baswaga chiefdoms, which, as noted above, were also devastated by raids and intergroup fighting during this period (Bergmans, 1953: 9).

25. Letter to the Commissaire de District Lubero from unidentified agent at Kiteranga, April 5, 1946, containing information from an interview with Kambale Milambo, an inhabitant of Kiteranga who witnessed Karakwenzi's activities. Seen in de Ryke Collection, Memorial Library, University of Wisconsin–Madison (46–6).

26. BHT 58, Kayingani Kambale, July 29, 1974.

27. In this context it is perhaps significant that while conflicts arising out of disputes between groups within Bashu society are frequently attributed to the refusal to give beer or violations of the etiquette of beer drinking, an activity that symbolizes commensality in Bashu traditions, none of the traditions relating to these external raids refers to beer drinking.

28. Kasumbakali subsequently sought refuge with his cousin Mbonzo, mwami of Malio, who recognized his authority and subsequently refused to acknowledge that of Vyogho. BHT 101, Savataki Muhindo, October 10, 1974; BHT 122, Nzeghi and Kasubaho, November 12, 1974; F. Van Rompaey, "Les Règles coutumières régissant la succession des chefs Bashu," p. 7, AZB.

29. BHT 51, Metya Kataka, July 19, 1974; BHT 53, Mutamo and Sabuni, July 24, 1974.

30. BHT 59, Paul Muhindo, August 3, 1974; BHT 69, Katafali et al., August 9, 1974; BHT 78, Mbugha Kahindo, August 14, 1974.

31. BHT 167, Kavwaro, February 15, 1975; BHT 53, Mutamo and Sabuni, July 19, 1974; BHT 78, Mbugha Kahindo, August 15, 1974.

32. I had originally assumed that references to this investiture in Bashu traditions were examples of the stereotypic formula (a chief was invested and ended a period of famine) that occurs frequently in Bashu traditions and reflects

the general associations of a *mwami*'s reign with the well-being of the land and of interregna with famine. A rereading of the traditions, however, revealed that the verb used to describe this event was not *-singa*, which is the normal term for the accession, but *-subya*, which means to renew or start again.

33. BHT 79, Saa II, August 15, 1974; BHT 67, Kahoma et al., August 7, 1974. In recent years the timing of the annual planting ceremonies throughout the Bashu region has apparently been altered again, to coincide with the European new year.

34. BHT 89, Kasoro, September 13, 1974; BHT 85, Rutava, August 28, 1974; BHT 86, Kaoma Kamate and Masekini Kasereka, August 30, 1974.

35. BHT 17, Kavota Mugasi, June 1, 1974.

36. BHT 123, Rutava, November 13, 1974.

37. BHT 43, Kitanda Viavo, July 6, 1974; BHT 31, Iteheri, June 24, 1974.

38. Evidence for Vyogho's acquisition of control over the distribution of *mikene* comes from the testimony of a *mukumu w'omukene*, BHT 57, Kasango, July 27, 1974, and is supported by the fact that *bakumu* in Malio and Maseki, where ritual centralization was not as strong, still maintain their control over the distribution of the *mikene*.

39. Thus, during the annual planting ceremonies these client chiefs were responsible for redistributing the *mikene* that they had received from Vyogho and for directing local ritual observances, though in each case the local ritual leaders played a major role.

40. BHT 17, Kavota Mughasi, June 1, 1974; BHT 53, Mutamo and Sabuni, July 24, 1974; BHT 83, Sabuni, August 26, 1974.

CHAPTER EIGHT: COLONIAL RULE
AND THE POLITICS OF RITUAL CHIEFSHIP

1. "Rapport sur la reconnaissance effectuée chez Lukanda et Kakuse, Août 1913," Registre des Renseignements Politiques-Kasindi 1906–1913, AZB. "Reconnaissance fait Octobre 1911," Registre des Renseignements Politiques-Kasindi, 1906–1913. "Among the chiefs Lukanda, Binga and Molenge, the inhabitants seemed disposed to come to the post. Everyone was in their huts during my passage. To the contrary at Tsombira's, Kakusa's and Bunda's, everyone had abandoned their huts. . . . In fact the inhabitants have declared to the messenger who preceded that they wish to have no relations with the whites."

2. The Bashu region, included at the time in the Territoire de la Semliki, was placed under military occupation by order no. 946 of April 17, 1922, issued by the Commissaire du Kivu, Van de Ghiste. For a more detailed reporting of the official account of the Ransbotijn expedition, see L. Bergmans (1974: 49).

3. An eyewitness described the incident as follows: "Kawaya [Bashu name for Ransbotijn] was attacked at Maghigi, but he in turn killed many people. He then fled towards Kasongwere [to the south]. Then he returned to Kyondo. Early in the morning we heard gunfire and fled to Lubwe. They killed many here including the son of Kalengeghya and the son of Mukatsi. We stayed two days at Lubwe. Then we heard Kawaya announce his departure with a trumpet. We returned and found many dead." BHT 67, Kaghoma et al., August 7, 1974. For other accounts, BHT 60, Kalenghegha Musakwa, August 4, 1974; BHT 74, Kinyandali, August 13, 1974.

4. The Belgians launched two expeditions into the southern Nande region prior to the First World War in 1908 and again in 1912. Both expeditions failed

to achieve their intended goal, which was the submission of the chiefs of this region. A primary source of resistance was Maboko, the son of Karakwenzi, who, following his father's arrest and exile, had returned to the Mitumbas and established a wide network of alliances in the areas south of the present Bashu region. The story of Maboko's activities in the mountains and of his resistance to Belgian incursions has yet to be fully researched. For a Belgian account of Maboko, see "Extraits de la rapport du Commissaire de District Hackers: Organisation de la population du Territoire de la Semliki," by F. Absil, 1927, in de Ryck Collection, Memorial Library, University of Wisconsin–Madison.

5. BHT 24, Kassemengo, June 12, 1974; "Rapport sur la reconnaissance effectuée chez Lukanda, Août 1913," Registre des Renseignements Politiques, Kasindi, 1908–1913; P. Joset, "Histoire des Beni, 1889–1938," p. 71. Kitawiti first presented himself in 1913 to a Belgian reconnaissance team and supplied them with food. Kitawiti's submission to Patefoort occurred five months before the submission of the rest of the chiefs of Isale and the southern Bashu region. Bergmans suggests that Kitawiti in fact encouraged other Bashu chiefs to resist Belgian overtures and then secretly submitted himself (Bergmans 1974: 53).

6. "Observations sur Biogo," Registre des Renseignements Politiques-Beni, 1908–1911, AZB.

7. F. Absil, "Organisation de la Population du Territoire de la Semliki," 1927, pp. 12–13, AZB.

8. Personal communication from Richard Sigwalt, who has conducted extensive research on Belgian patterns of political expansion in the Kivu region and particularly among the Bashi of South Kivu. The 'grand chef' policy of indirect rule appears to have originated with Louis Frank, who became Belgian Colonial Minister in 1917.

9. F. van Rompaey noted that Kitawiti, continuing his father's politics, ". . . maintained and reinforced the relations which his father had created abroad. In return for his services, offerings were no longer volunteered, but were demanded by him. From all regions (from the Baswaga, Batangi and the Bamate in the present territory of Lubero and from a part of Rutshuru) he received gifts which began to resemble tribute. He is reputed to have been able to send clouds of locusts to those who resisted his demands." "Les Règles coutumières régissant la succession des chefs Bashu," February 10, 1945, p. 8, AZB. The last claim was repeated by one of my own informants, BHT 28, Mombokani Wasokundi, June 17, 1974.

10. Report to the Territorial Administrator-Semliki, June 1, 1925 (283/Pol./B), unsigned. Kitawiti is said to have returned from Beni on May 8, 1925. On the night of the tenth and eleventh, there was a celebration of his return, at which he drank some beer and shortly thereafter died. The makers of the beer fled upon hearing the news. This led some people to assume that the beer had been poisoned.

11. Mukanero is reported to have refused to give any aid to the local Belgian administrator in collecting taxes and organizing work forces. H. Coune à Commissaire de District, Beni, 18 November 1929 (725/Pol./C.), AZB.

12. Correspondence: H. Daniels à Commissaire de District, Beni, August 19, 1927 (123/Pol. Bashu), AZB.

13. Correspondence: Coune à Commissaire de District, Beni, November 18, 1929.

14. Correspondence: Coune à Personel Européen en Occupation, August 26, 1931 (Pol. c/ no. 44); R. Moriame à l'Administrateur Territorial, Beni, August

15, 1923 (369/Doss. Buyora); de Schampheleer à Commissaire de District, August 23, 1932, AZB; Joset (1939: 73).

15. Correspondence: H. Daniels à H. Henne, August 24, 1927; H. Daniels à Commissaire de District Irumu, August 19, 1927 (123/Pol. Bashu); Daniels à Commissaire de District Irumu, August 26, 1927 (493/Pol. Bashu); F. Absil à l'Administrateur Territorial Beni, November 16, 1930, AZB.

16. Belgian administrative reports before 1932 are consistent in ascribing the office of *grand chef* to Kitawiti and his descendants. "Rapport d'enquête sur l'Isale de Kitawite (1923)," Registre des Renseignements Politiques Semliki, 1927–29, pp. 6–13; "Rapport d'enquête sur la chefferie des Bashu," Registre des Renseignements Politiques-Beni, 1928–37. A second report on the Isale of Kitawiti included within the Registre des Renseignements Politiques-Beni 1928–37 presents the revised view, which supports Muhayirwa's legitimacy. This report is entitled "Rapport d'enquête sur la *sous-chefferie* de Kitawiti." One source for this revised history is suggested by a letter from de Schampheleer to the Territorial Agent, Beni, August 23, 1932. Schampheleer claims that Vehamba, the *mwami* of Maseki who had been a primary leader in the campaign of passive resistance against Musayi, told him that he had rebelled against Musayi because Musayi ". . . is not our chief, our chief is Kamabo (Muhayirwa)." In a subsequent letter dated August 24, 1932, de Schampheleer comments on this claim, stating that "I have no confidence in the declaration of Weamba, who is not exactly friendly with Musayi, but I have observed that other notables as well as the *capitas* of Weamba's group show a certain deference to Muhayirwa." It is, of course, possible that Schampheleer fabricated this evidence to justify replacing Musayi. This is suggested by the fact that in the same letter (August 24, 1932), he claims that Musayi himself acknowledged Muhayirwa's authority, and that according to Musayi, Kitawiti had been invested because Muhayirwa was too young. This claim seems highly unlikely, given what is known about Musayi's character. On the other hand, de Schampheleer may simply have been strengthening his case, which was, in fact, based on Bashu testimonies.

17. Van Rompaey, who was Territorial Administrator at the time, indicated that Muhayirwa's downfall was caused by the discontent of other chiefly lines at having their power and independence sublimated to that of a single chief. "Règles coutumière régissant la succession des chefs Bashu," p. 8.

18. Labor demands for road construction were particularly oppressive to the Bashu because of the mountainous terrain over which the roads had to be built. Several Bashu elders commented that one of the main incentives for fleeing consolidated villages was to avoid these labor demands. Labor requirements could be also avoided by paying an exemption tax of 30 fr. Few Bashu cultivators could afford this outlay, however. Réunions de Conseil de la Chefferie des Bashu, December 28, 1931, AZB.

19. In 1935 the price paid for wheat, which was introduced by the Belgians, was .40 fr./kilo, and an additional .02fr./kilo was paid to the chief as a premium. Conseil de la Chefferie des Bashu, May 20, 1935, AZB.

20. Décision 11/32 Hygiène du 8 Juin 1932.

21. de Schampheleer à l'agent territorial, October 21, 1932, AZB.

22. Réunion du Conseil de la Chefferie des Bashu, May 20, 1935, AZB.

23. The Bashu continued to travel to Katwe illegally, on a reduced scale. Belgian administrators attempted to prevent these violations; however, it was impossible to do so given the manpower available for patrolling the valley. An anonymous letter to the Territorial Agent Offenheim on July 25, 1934, recom-

mended that an official route be established between the Bashu region and Katwe, since it was impossible to prevent traders from crossing the valley.

24. Joset (1939: 22). The Minière des Grandes Lacs evidently reduced their expenses by buying their food supplies closer at hand, thereby cutting transport costs involved in buying produce among the Bashu. Correspondence from de Schampheleer, August 23, 1932, AZB.

25. BHT 7, Mombokani Wasokundi, March 23, 1974; BHT 8, Mombokani Wasokundi, March 27, 1974; BHT 166, Mwendokolero, February 14, 1975.

26. BHT 163, Kasogho Kambale, February 11, 1975.

27. R. Moriame à Commissaire de District, February 3, 1932; R. Moriame à l'Administrateur Territorial de la Semliki, December 27, 1931.

28. A.I.M.O. Rapport, District du Kivu, 1941, seen in de Ryck Collection, Memorial Library, University of Wisconsin–Madison, 48–1.

29. For a more complete analysis of changing concepts of misfortune among the Bashu during the colonial period, see R. Packard, 1980b.

30. BHT 93, Kasoro, September 21, 1974.

31. BHT 55, Mombokani Wasokundi, July 24, 1974.

32. BHT 45, Kaghoma Vighetsi Mughashu, July 13, 1974; BHT 133, Kisuki, November 26, 1974.

33. Correspondence: F. Absil à Administrateur Territorial de Beni, September 17, 1935; Absil à l'Administrateur Territorial de Beni, October 7, 1935; H. Preumont, "Propositions en vue de la destitution comme chef de tous les Bashu Chef actuel Muhayirwa et en vue de son Remplacement par le notable Muhashu," September 28, 1935. In essence, the Belgian records returned to the pre-1932 formula, which indicated that Luvango's descendants were the true *grands chefs* of the Bashu.

34. Correspondence: F. van Rompaey à l'Administrateur Territorial de Beni, October 27, 1944 (80/AIMO/B.), AZB. Réunion du Conseil de la Chefferie Bashu, July 22 and 23, 1936, AZB. Muhashu is accused of allowing Bashu to return to their homes outside the consolidated villages.

35. Réunions du conseil de la chefferie Bashu, May 10–13, 1943, AZB.

36. Réunions du conseil de la chefferie Bashu, August 9, 1939, AZB.

37. A.I.M.O. Rapport, Kivu District, 1945, seen in de Ryck Collection, Memorial Library, University of Wisconsin–Madison, 48–4. The actual percentage in 1945 was 34.47. This was up from 32.59 percent at the end of 1944.

38. Bashu resistance to soybean cultivation took several forms, including sorcery, cutting the roots of the plants so that they would appear to be unsuitable for the area, and simply refusing to plant them. Katembo (1974: 68–69).

39. A.I.M.O. Report, Kivu District, 1943, de Ryck Collection, 48–2.

40. Correspondence: F. van Rompaey à l'Administrateur Territorial de Beni, November 11, 1936. P. E. Joset, "Procès-Verbal: Réunion des notables Wasongora, Kasindi, November 27–29, 1936," AZB.

41. Réunion du Conseil de la chefferie Bashu, May 20, 1935, AZB.

42. Katembo (1974: 68) suggests that the agricultural demands of Belgian administrators, and particularly the forced cultivation of soybeans, was a cause for many men to seek work in the mines or to migrate to Uganda.

43. For further details of this development, see Packard, 1980b.

44. A.I.M.O. Report, Kivu District, 1943, de Ryck Collection, 48–2; Réunion du Conseil de la Chefferie Bashu, May 10–13, 1943, AZB. The extended drought caused many Bashu cultivators simply to abandon their fields.

45. F. van Rompaey, "Les Règles coutumières régissant la succession des chefs Bashu," February 10, 1945, p. 11, AZB. BHT 7, Mombokani Wasokundi, March 26, 1974; BHT 29, Rutava, June 19, 1974; BHT 43, Kitanda Viavo, July 6, 1974.

46. BHT 29, Rutava, July 19, 1974.

47. Correspondence: F. van Rompaey à l'Administrateur Territorial de Beni, October 27, 1944 (80/AIMO/B.), AZB.

48. Réunion du Conseil de la Chefferie Bashu, March 28, 1945; Correspondence: F. van Rompaey à l'Administrateur Territorial de Beni, February 13, 1945; van Rompaey à l'Administrateur Territorial de Beni, March 11, 1945 (36/AIMO/B), AZB.

GLOSSARY

bakaka	persons responsible for preparing the body of a deceased *mwami* for burial
basingya	council of investors, responsible for selecting a new *mwami* and for overseeing his investiture
bakonde	'clearers of the forest' and their descendants; play an important role in the performance of rituals designed to protect the land and ensure its productivity
basumba	forest dwellers who were former occupants of the Bashu region, regarded by the Bashu as possessing special ritual influence over the world of the forest
mbita	royal diadem fabricated out of the skin of a flying squirrel, which is covered with a cap made of knotted raffia fibers, leaves of wild plants, and the teeth of wild animals
mikene	ritually powerful seeds thought to ensure the productivity of gardens in which they are placed
mombo	ritual wife and female counterpart of the *mwami*, responsible for 'producing' the *mwami*'s successor; represents the chiefship during interregna
mughula	charged with removing and guarding the jawbone of a deceased *mwami*
muhako	a payment of several goats given annually by members of a localized lineage to the group that gave them land. All Bashu pay *muhako* to their *mwami* in recognition of his 'ownership' of all land within his chiefdom.
muhangami	protector of the *mwami*, responsible for ensuring that the *mwami* is protected from pollution
mukulu	senior member and representative of a lineage *(nda)*, responsible for sacrifices to lineage ancestors; *mukulu* of royal lineage often called *semwami* or *muhito*
mukumu	a healer who treats people
mukumu w'ekihugho	a healer who works with the land in order to ensure its productivity
musingya	1. a member of the *basingya*; 2. the person who presents the *mwami* with his *mbita*
mwami w'embita	invested chief and central ritual leader of a Bashu chiefdom
ngemu	tribute in the form of a portion of one's harvest, given to ritual leaders, including the *mwami*, who have through their ritual actions ensured the productivity of one's gardens
semombo	title given to individual who provides the *mwami* with his *mombo*, literally 'father of the mombo,' though the

228

	semombo is not necessarily the biological father of the *mombo*.
semwami	see *mukulu*
vighala	grave within which the *mwami* spends the night with the *omusumbakali,* a ritual wife, during his accession rites
vuhere vw'ovusyano	annual new year rite directed by the *mwami w'embita,* held prior to the planting of eleusine in September

REFERENCES

ARCHIVAL SOURCES

Several archival collections were consulted for this study. In Zaire, the zone archives at Beni (AZB) and Lubero (AZL) proved quite valuable, despite the fact that some documents had been lost during the political troubles of 1964. Most of the important *Registres des Renseignements Politiques* were available for both zones, though the Registre for the Baswaga collectivity had disappeared from Lubero. Of particular importance for this study were the *Registres des Renseignements Politiques-Kasindi, 1908–1913,* which contained data collected from among the population of the upper Semliki Valley prior to the evacuation of the valley on account of sleeping sickness in the early 1930s. In the absence of informants presently living in the plains, these reports provided valuable information on the relationship that existed between the former occupants of the plains and the Bashu. Also of considerable interest at Beni were the *Rapports des Reunions du Conseil de la Chefferie Bashu*. These contained important data on land and succession disputes, which in turn contained valuable historical information. Finally, a number of ethnographic reports exist at both Beni and Lubero. The most valuable of these were the reports on Banande culture by F. Absil in 1923 and F. van Rompaey's reports on Bashu succession rules, written in the 1930s and 1940s.

The administrative centers of the various Nande collectivities (former chefferies) also contain some archival materials. Unfortunately, the Bashu records at Vuhovi were almost totally destroyed during the 1964 rebellion. While I was in Zaire an attempt was being made to obtain copies of letters sent to and from Vuhovi before the rebellion.

Outside Zaire, the archival materials possessed by the ethnographic section of the Musée Royal de l'Afrique Centrale in Tervuren, Belgium, were very helpful, as were the excellent bibliographic resources. Of particular interest were the Dossiers du Kivu-Nord, which contain a number of ethnographic studies written by colonial agents and administrators in Zaire, as well as P. E. Joset's informative "Historique du Territoire de Beni 1889–1938."

Material was also consulted at the Archives Africains, Place Royal, Brussels. However, the strictly enforced fifty-year rule prevented my gaining access to reports and correspondence written after 1923, the year in which the Belgians first occupied the Bashu chiefdoms.

The de Ryck Collection located at the Memorial Library, University of Wisconsin, contained important data on the activities of Karakwenzi among the Bashu, as well as several A.I.M.O. reports for the 1940s that were not available elsewhere. Finally, British Parliamentary Papers consulted at Widener Library, Harvard University, contained information on events that occurred in southwest Uganda during the 1890s. Of special interest were the "Papers Relating to the Execution of Mr. Stokes in the Congo State," Parliamentary Papers, 1896, LIX, which provided data on the activities of the British ivory trader Carl Stokes, the ivory trade, and the spread of firearms in the upper Semliki Valley.

WORKS CITED

Alpers, Edward. 1969. "Trade, State and Society among the Yao in the Nineteenth Century." *Journal of African History* 10 (3): 405–20.
Alube, L. 1934. "Le Droit coutumier chez les Bashu." *Bulletin Cercle Colon. Luxembourg* 4: 10–19.
Avua, L. 1968. "Drought Making among the Lugbara." *Uganda Journal* 32 (1): 29–38.
Bailey, F. G. 1969. *Strategems and Spoils.* New York: Schocken Books.
Baitwababo, S. R. 1972. "Foundations of Rujumbura Society." In *A History of Kigezi,* ed. D. Denoon. Kampala: Adult Education Bureau.
Baxter, P. T. W., and Butt, Audrey. 1953. *The Azande and Related Peoples.* London: International Africa Institute.
Beattie, John. 1972. *The Nyoro State.* Oxford: Clarendon Press.
———. 1964. "Rainmaking in Bunyoro." *Man* 64: 140–41.
Beidelman, T. O. 1970. "Myth, Legend and Oral History." *Anthropos* 65 (5–6): 74–97.
———. 1966. "Swazi Royal Ritual." *Africa* 36: 373–405.
Berger, Iris. 1980. *Religion and Resistance: East African Kingdoms in the Precolonial Period.* Tervuren: Musée Royal de l'Afrique Centrale.
Bergmans, Lieven. 1974. *Histoire des Bashu.* Butembo: Editions A. B. B.
———. 1971. *Les Wanande: croyances et pratiques traditionelles.* Butembo: Editions A.B.B.
———. 1970. *L'Histoire des Baswaga.* Butembo: Editions A.B.B.
———. 1958. "Rogatiens paiennes." *Afrique Ardent* 39 (105): 14–16.
———. 1953. "Les Chevres chez les Baswaga." *Afrique Ardent* 18 (74): 8–12.
———. 1952. "Éclair et foudre chez les Wanande." *Afrique Ardent* 17 (72): 14–18.
Biebuyck, Daniel. 1979. *Hero and Chief.* Berkeley: University of California Press.
———. 1966. *Rights in Land and Resources among the Nyanga.* Brussels: Academie des Sciences des Outre-Mer.
———. 1956. "Organisation politique des Nyanga: la chefferie Ihana." *Kongo-Overzee* 22 (4–5): 301–41; 23 (1–2): 58–98.
Biebuyck, Daniel, and Mateene, Kahombo. 1971. *The Mwindo Epic.* Berkeley: University of California Press.
Borgerhoff, F. 1912. "Les Industries des Wanande." *La Revue Congolaise* 3: 278–84.
Brain, James L. 1973. "Ancestors as Elders: Some Further Thoughts." *Africa* 43 (1): 122–33.
Buchanan, Carole. 1974. "The Kitara Complex: The Historical Tradition of Western Uganda to the Sixteenth Century." Ph.D. dissertation, Indiana University.
Charsley, S. R. 1969. *The Princes of Nyakyusa.* Nairobi: East African Publishing House.
Chretien, J.-P. 1977. *La Royauté capture les rois.* Nanterre: Laboratoire d'Ethnologie et de Sociologie Comparative.
Claessens, J. 1929. "Du lac Albert au lac Kivu, à travers les regions montagneuses longéant la frontière orientale de la colonnie." *Bulletin Agricole du Congo Belge* 20 (1): 4–56.
Cohen, David W. 1977. *Womunafu's Bunafu.* Princeton: Princeton University Press.

————. 1974. "A Preliminary Survey of Climatic Trends in the Lake Region of East Africa." ASA Conference, Chicago.

————. 1970. "A Survey of Interlacustrine Chronology." *Journal of African History* 11 (2): 177–202.

Colle, Pierre. 1921. "L'Organisation politique des Bashi." *Congo* 2 (5): 657–84.

Curtin, P.; Feierman, S.; Thompson, L.; and Vansina, J. 1977. *African History*. Boston: Little, Brown.

Debatty, R. 1951. "Étude des coutumes des Wanande en territoire Lubero: évolution du pouvoir ou investiture du 'mwami'." *Bulletin des Juridictions Indigénes* 19 (6): 174–88.

de Heusch, Luc. 1966. *Rwanda et la civilisation interlacustrine*. Brussels: Universite Libré de Bruxelles.

————. 1958. *Essai sur le symbolisme de l'incest royal en Afrique*. Brussels: Universite Libré de Bruxelles.

Denoon, Donald, ed. 1972. *A History of Kigezi in Southwest Uganda*. Kampala: Adult Education Bureau.

d'Hertefelt, Marcel. 1971. *Les Clans du Rwanda ancien*. Tervuren: Musée Royal de l'Afrique Centrale.

d'Hertefelt, Marcel, and Coupez, A. 1964. *La Royauté sacrée de l'ancien Rwanda*. Tervuren: Musée Royal de l'Afrique Centrale.

Douglas, Mary. 1973. *Natural Symbols*. New York: Penguin.

————. 1966. *Purity and Danger*. London: Routledge and Kegan Paul.

Edel, M. 1968. *The Chiga of Western Uganda*. London: International Africa Institute.

Ehret, Christopher. 1974. "Some Thoughts on the Early History of the Nile-Congo Watershed." *Ufahamu* 2: 85–122.

Evans-Pritchard, E. E. 1972. *The Azande: History and Political Institutions*. Oxford: Clarendon Press.

————. 1962. "Divine Kingship of the Shilluk of the Nilotic Sudan." In *Essays on Social Anthropology*. London: Oxford University Press.

————. 1956. *Nuer Religion*. London: Oxford University, Press.

————. 1940. *The Nuer*. London: Oxford University Press.

————. 1937. *Witchcraft, Oracles and Magic among the Azande*. Oxford: Clarendon Press.

Fallers, L. A. 1965. *Bantu Bureaucracy*. Chicago: University of Chicago Press.

Feierman, Steven. 1974. *The Shambaa Kingdom*. Madison: University of Wisconsin Press.

————. 1972. "Concepts of Sovereignty among the Shambaa and their Relation to Political Action." Ph.D. dissertation, Oxford University.

Ford, John. 1979. "Ideas which have influenced attempts to solve the problem of African Trypanosomiasis." *Social Science and Medicine* 13b (4): 269–76.

————. 1971. *The Role of the Trypanosomiasis in African Ecology*. Oxford: Clarendon Press.

Ford, John, and Hall, R. de Z. 1947. "History of Karagwe." *Tanganyika Notes and Records* 24: 3–27.

Fortes, Meyer. 1940. "The Political Systems of the Tallensi of the Northern Territories of the Gold Coast." In *African Political Systems*, ed. M. Fortes and E. E. Evans-Pritchard. London: International Africa Institute.

Fraas, Pauline. 1961. *A Nande-English, English-Nande Dictionary*. Washington, D.C.

Freedman, Jim. 1974. "Ritual and History: The Case of Nyabingi." *Cahiers d'É-tudes Africaines* 14 (53): 170–80.

———. 1979. "Three Murari's, Three Gahaya's and the Four Phases of Nya-vingi." In *Chronology, Migration and Drought in Interlacustrine Africa*, ed. J. B. Webster. New York: Africana Publishing.

Geraud, F. 1972a. "The Settlement of the Bakiga." In *A History of Kigezi in Southwest Uganda*, ed. D. Denoon. Kampala: Adult Education Bureau.

———. 1972b. "Historical Notes on the Bakiga of Uganda." *Annali del Ponticio Museo Missionario Ethnologico* 34: 293–357.

Gluckman, M. 1954. "Succession and Civil War among the Bemba." *Journal of the Rhodes-Livingstone Institute* 6: 6–25.

Good, Charles M. 1972. "Salt Trade and Disease: Aspects of Development in Africa's Northern Great Lakes Region." *International Journal of African Historical Studies* 5 (4): 543–86.

Goody, Jack., ed. 1966. *Succession to High Office*. Cambridge Papers in An-thropology. Cambridge: Cambridge University Press.

Gorju, P. Julien. 1920. *Entre le Victoria, l'Albert et l'Edouard*. Rennes: Imprimeries Oberthur.

Gray, Richard, and Birmingham, David. 1970. *Precolonial African Trade: Essays on Trade in Central and Eastern Africa*. London: Oxford University Press.

Hartwig, Gerald. 1976. *The Art of Survival in East Africa*. New York: Africana Publishing.

Heinzelin, Jean de. 1957. *Les Fouilles d'Ishango*. Brussels: Institut des Parcs Na-tionaux du Congo Belge.

Henige, David. 1974. *The Chronology of Oral Tradition: Quest for a Chimera*. Oxford: Clarendon Press.

Herring, Robert S. 1979. "Hydrology and Chronology: The Rodah Nilometer as an Aid to Interlacustrine Chronology." In *Chronology, Migration and Drought in Interlacustrine Africa*, ed. J. B. Webster. New York: Africana Publishing.

Ingham, Kenneth, 1975. *The Kingdom of Toro in Uganda*. London: Methuen and Co.

James, Wendy. 1972. "The Politics of Rain Control in Uduk." In *Essays in Sudan Ethnography*, ed. W. James and I. Cunnison. London: C. Hurst.

Johnston, Sir. H. 1904. *The Uganda Protectorate*. London: Hutchinson and Co.

Jones, D. H. 1970. "Problems of African Chronology." *Journal of African History* 11 (2): 161–77.

Joset, P. E. 1939. "Historique du territoire de Beni, 1889–1938." Typescript. Dossiers du Kivu Nord, Section Ethnographique, Musée Royal de l'Afrique Centrale.

K. W. 1937. "The Procedure in Accession to the Throne of a Nominated King in the Kingdom of Bunyoro-Kitara." *Uganda Journal* 3 (2): 289–99.

———. 1935–37. "The Kings of Bunyoro-Kitara." *Uganda Journal* 3 (2): 149–60; 4 (1): 65–83; 5 (1): 53–92.

Kagame, Alexis. 1972. *Un Abrege de ethno-historie du Rwanda*. 2 vols. Butare: Editions Universitaire du Rwanda.

———. 1963. *Les Milices du Rwanda précolonial*. Brussels: Academie des Sciences d'Outre-Mer.

Kaggwa, Apolo. 1971. *The Kings of Buganda*. Translated by M. Kiwanuka. Nairobi: Longmans.

———. 1934. *Empisa za Baganda*. New York: Columbia University Press.

Kahiwa, Ndwatwa Kasereka. 1975. "Evolution Politico-Administrative de la Col-
    lectivité Bashu." Memoire, Faculté des Sciences Sociales, Administratives et
    Politiques, UNAZA, Lubumbashi.
Kakiranyi, Kule. 1972. "Le couronnement du mwami dans la tradition Nande."
    Memoire de License, I.S.P. Bukavu.
Kamuhangire, E. R. 1975. "The Precolonial Economic and Social History of East
    Africa with Special Reference to South-Western Uganda Salt Lakes Re-
    gion." In *Hadith 5*, ed. B. A. Ogot. Nairobi: East African Publishing House.
————. 1972. "The States of Southwestern Uganda Salt Lakes Region During
    the Nineteenth Century." Seminar Paper, Department of History, Makerere
    University. Kampala.
Karp, Ivan. 1978. *Fields of Change among the Iteso of Kenya*. London: Routledge
    and Kegan Paul.
————. 1980. "Beer Drinking and Social Experience in an African Society: An
    Essay in Formal Sociology." In *Explorations in African Systems of Thought*, ed.
    I. Karp and C. Bird. Bloomington: Indiana University Press.
Karugire, S. R. 1972. *A History of the Kingdom of Nkore in Western Uganda to 1895*.
    Oxford: Clarendon Press.
Kataliko, Emmanuel. 1964. "Les Croyances traditionelles Nande." *Ann. Pontif.
    Museo Miss. Ethnol.* 28: 75–84.
Katembo, Kyoswire Kahayi. 1974. "La Résistance passive du Munande face à
    l'action coloniale en agriculture." Memoire de License, I.S.P. Bukavu.
Kiwanuka, M. S. M. 1971. *A History of Buganda*. London: Longmans.
Kjekshus, Helge. 1977. *Ecology Control and Economic Development in East Africa*.
    Berkeley: University of California Press.
Kopytoff, Igor. 1971. "Ancestors as Elders in Africa." *Africa* 41 (2): 129–41.
Kuper, Hilda. 1947. *A Pastoral Aristocracy*. London: Oxford University Press.
Lanning, E. C. 1954. "Masaka Hill: An Ancient Center of Worship." *Uganda
    Journal* 18 (1): 24–30.
Leach, Edmund. 1968. *Political Systems of Highland Burma*. Boston: Beacon.
Lindi, General Henri de la. 1948. "Historique sommaire de la campagne de la
    Lindi." *Bulletin de l'Institut Royal Colonial Belge* 2: 411–64.
Lugard, Lord Frederick. 1959. *The Diaries of Lord Lugard*. Vol. II, ed. M. Perham.
    Evanston: Northwestern University Press.
————. 1893. *The Rise of Our East African Empire*. Edinburgh: Blackwood.
Matte, J. S. In press. "The Bakonzo Conquest of the Ruwenzoris." In *Uganda
    Before 1900*. Vol. 1., ed. J. B. Webster. Nairobi: East African Publishing
    House.
Miller, Joseph C. 1974. *Kings and Kinsmen*. Oxford: Clarendon Press.
Moeller, A. 1936. *Les Grandes Lignes des migrations des Bantous de la province
    orientale du Congo belge*. Brussels: Institut Royal Colonial Belge.
Mworoha, Emile. 1977. *Peuples et rois de l'Afrique des lacs*. Paris: Nouv. Edns.
    Africaines.
Nicholson, Sharon E. 1976. "A Climatic Chronology for Africa: Synthesis of
    Geological, Historical and Methodological Information and Data." Ph.D.
    dissertation, University of Wisconsin–Madison.
Nicolet, J. 1972. "Regions qui se détacherent du Kitara et devennent des
    royaumes independent." *Annali Del Pontificio Museo Missionario Ethnologico*
    34: 227–92.
Nyakatura, J. W. 1973. *The Anatomy of an African Kingdom*. Hamden, Ct.:
    Archon.

Ogot, B. A. 1964. "Kingship and Statelessness among the Nilotes." In *The Historian in Tropical Africa*, ed. J. Vansina, R. Mauny, and L. V. Thomas. London: International Africa Institute.

Oliver, Roland. 1954. "The Baganda and the Bakonjo." *Uganda Journal* 18 (1): 31–33.

———. 1959. "Ancient Capital Sites of Ankole." *Uganda Journal* 23 (1): 51–63.

Oliver, Roland, and Matthews, Gervase, eds. 1963. *Oxford History of East Africa*. Oxford: Clarendon Press.

Packard, Randall. 1980a. "The Study of Historical Process in African Traditions of Genesis: The Bashu Myth of Muhiyi." In *The African Past Speaks*, ed. Joseph C. Miller. London: Dawson Publishing.

———. 1980b. "Social Change and the History of Misfortune among the Bashu of Eastern Zaire." In *Explorations in African Systems of Thought*, ed. I. Karp and C. Bird. Bloomington: Indiana University Press.

———. 1977. "Debating in a Common Idiom: Traditions of Genesis as Instruments of Political Integration in Traditional African States." ASA Conference, Houston.

Page, Melvin E. 1974. "The Manyema Hordes of Tippu Tip: A Case Study in Social Stratification and the Slave Trade in Eastern Africa." *International Journal of African Historical Studies* 7 (1); 69–84.

Pages, A. 1933. *Un Royaume hamite au centre de l'Afrique*, Brussels: Academie Royale de Belgique.

Park, G. K. 1966. "Kinga Priests: The Politics of Pestilence." In *Political Anthropology*, ed. Mark Swartz, Victor Turner, and Arthur Tuden. Chicago: Aldine.

Pasha, Emin (Edouard Schnitzer). 1888. *Emin Pasha in Central Africa*. Translated by G. Scheinfurth. London.

Posnansky, Merrick. 1966. "Kingship, Archeology and Historical Myth." *Uganda Journal* 30 (1); 1–12.

Rappaport, Roy A. 1969. *Pigs for the Ancestors*. New Haven: Yale University Press.

Ray, Benjamin. 1976. *African Religions*. Englewood Cliffs, N.J.: Prentice-Hall.

Richards, Audrey I. 1961. "African Kings and their Royal Relatives." *Journal of The Royal Anthropological Institute* 91: 135–49.

———. 1960. "Social Mechanisms for the Transfer of Political Rights." *Journal of the Royal Anthropological Institute* 90 (2): 175–90.

———. 1959. *East African Chiefs*. New York: Faber.

Riehl, Herbert, and Meitin, Jose. 1979. "Discharge of the Nile River: A Barometer of Short Term Climatic Variation." *Science* 206: 1178–80.

Rigby, Peter. 1968. "Gogo Rituals of Purification." In *Dialectic in Practical Religion*, ed. E. R. Leach. Cambridge: Cambridge University Press.

Roberts, Andrew. 1973. *A History of the Bemba*. Madison: University of Wisconsin Press.

———. 1969. "Political Change in the 19th Century." In *A History of Tanzania*, ed. I. Kimambo. Nairobi: East African Publishing House.

Roscoe, John. 1966. *The Baganda*. New York: Macmillan.

———. 1923. *The Bakitara or Banyoro*. Cambridge: Cambridge University Press.

———. 1923. *The Banyankole*. Cambridge: Cambridge University Press.

Rukidi III, Sir George Kamurasi. N.d. "The Kings of Toro." Typescript. Translated by Joseph R. Muchope. History Department, Makerere University, Kampala.

Rwankwenda, M. M. R. 1972. "A History of Kayonza: The History of the Ruling Dynasty." In *A History of Kigezi in South-West Uganda*, ed. D. Denoon. Kampala: Adult Education Bureau.

Sahlins, Marshall. 1963. "Poor Man, Rich Man, Big Man, Chief: Political Types in Melanesia and Polynesia." *Comparative Studies in Society and History* 5 (3): 285–303.

Schebesta, Paul. 1936. *My Pygmy and Negro Hosts.* London: Hutchinson.

Schecter, Robert. 1976. "History and Historiography on a Frontier of Lunda Expansion." Ph.D. dissertation, University of Wisconsin–Madison.

Scott Eliot, G. F. 1896. *A Naturalist in Mid-Africa.* London: Innes and Co.

Seligman, C. G. 1932. *Pagan Tribes of the Nilotic Sudan.* London: G. Routledge and Co.

Shorter, Aylward. 1972. *Chiefship in Western Tanzania.* Oxford: Clarendon Press.

Sigwalt, Richard. 1975a. "The Early History of Rwanda: The Contribution of Comparative Ethnography." *History in Africa* 2: 137–46.

———. 1975b. "The Early History of Bushi: An Essay on the Historical Use of Genesis Traditions." Ph.D. dissertation, University of Wisconsin–Madison.

Southall, A. W. 1953. *Alur Society.* Cambridge: W. Hefferand and Sons.

Southwold, Martin. 1966. "Succession to the Throne of Buganda." In *Succession to High Office*, ed. Jack Goody. Cambridge: Cambridge University Press.

———. 1961. *Bureaucracy and Chiefship in Buganda.* East African Studies No. 14. Kampala: East African Institute of Social Research.

Stanley, H. M. 1890. *In Darkest Africa.* Vol. 2. New York: Charles Scribner and Sons.

———. 1889. "Letters From . . . " *Proceedings of the Royal Geographical Society of London* 11: 724–25.

Steinhart, E. 1979. "The Kingdoms on the March: Speculation on Social and Political Change." In *Chronology, Migration and Drought in Interlacustrine Africa*, ed. J. B. Webster. New York: Africana Publishing.

Stuhlman, F. 1894. *Mit Emin Pascha ins Herz von Afrika.* Berlin: Dietrich Reimer.

Swartz, Mark; Turner, Victor; and Tuden, Arthur. 1966. *Political Anthropology.* Chicago: Aldine.

Taylor, Brian. 1962. *Western Lacustrine Bantu.* Ethnographic Survey of Africa. London: International Africa Institute.

Thiry, E. 1963. "Bahima et migrations des Lwo." *Africa-Tervuren* 9 (4): 105–107.

Tosh, John. 1970. "The Northern Interlacustrine Region." In *Precolonial African Trade*, ed. R. Gray and D. Birmingham. London: Oxford University Press.

Trevor-Roper, H. R. 1969. "The European Witchcraze of the Sixteenth and Seventeenth Centuries." In *The European Witchcraze and Other Essays.* New York: Harper and Row.

Tripe, W. B. 1939. "The Death and Displacement of a Divine King in Uha." *Man* 39 (21): 22–25.

Tucker, A. N. 1960. "Notes on Konzo." *African Language Studies* 1: 16–24.

Turner, Victor. 1967. *Forest of Symbols.* Chicago: University of Chicago Press.

Uzoigwe, G. N. 1973. "Succession and Civil War in Bunyoro-Kitara." *International Journal of African Historical Studies* 6 (1): 49–71.

———. 1970. "Kabarega and the Making of a New Kitara." *Tarikh* 3 (2): 5–21.

Van Geluwe, H. 1956. *Les Bira et les peuplades limitrophes.* Tervuren: Musée Royal de l'Afrique Centrale.

Vansina, Jan. 1978a. *The Children of Woot.* Madison: University of Wisconsin Press.

———. 1978b. Review of *l'Intronisation d'un mwami,* by Pascal Ndayishinguje, and *La Royaute capture les rois,* by C.-P. Chretien. *International Journal of African Historical Studies* 11 (2): 327–28.

———. 1973. "L'Influence du mode de comprehension historique d'une civilisation sur les traditions d'origine: l'exemple Kuba." *Bulletin de l'Academie Royale des Sciences d'Outre-Mer* 2: 220–40.

———. 1962a. *L'Evolution du royaume Rwanda des origines à 1900.* Brussels: Academie Royale des Sciences d'Outre-Mer.

———. 1962b. "A Comparison of African Kingdoms." *Africa* 32 (4): 324–35.

Verdonck, H. 1928. "Déces du Mwami Rushambo et intronisation du Mwami Bahole, District du Kivu, Territoire du Buhavu." *Congo* 1 (3): 294–309.

Vervolet, Lt. G. 1909. "Aux sources du Nil." *Bulletin de la Societé Royale Belge de Geographie* 33 (1): 253–98, 395–430.

Viaene, L. 1952a. "La Religion des Bahunde, *Kongo-Overzee* 18 (5): 388–425.

———. 1952b. "L'Organisation politique des Bahunde." *Kongo-Overzee* 18: 8–34, 111–21.

———. 1951. "La Vie domestique des Bahunde." *Kongo-Overzee* 17 (2): 111–56.

Wachsmann, K. P., and Trowell, M. 1953. *Tribal Crafts of Uganda.* London: Oxford University Press.

Wayland, E. J. 1934. "Katwe." *Uganda Journal* 1 (2): 96–106.

Weber, Max. 1958. *From Max Weber.* Ed. H. H. Gerth and C. Wright Mills. New York: Oxford University Press.

Webster, J. B. 1979. "Noi!, Noi!, Famines as an Aid to Interlacustrine Chronology." In *Chronology, Migrations and Drought in Interlacustrine Africa,* ed. J. B. Webster. New York: Africana Publishing.

Williams, F. Lukyn. 1938. "Hima Cattle." *Uganda Journal* 6 (1): 17–24; (2): 87–117.

———. 1937. "The Inauguration of the Omugabe of Ankole to Office." *Uganda Journal* 4 (4): 300–301.

Wilson, Monica. 1959. "Divine Kingship and the 'Breath of Men'." Fraser Lecture. Cambridge: Cambridge University Press.

Winter, E. 1956. *Bwamba.* Cambridge: Cambridge University Press.

Wollaston, A. F. R. 1908. *From Ruwenzori to the Congo.* New York.

Wrigley, Christopher C. 1973. "The Story of Rukidi." *Africa* 43: 219–34.

Young, Michael W. 1966. "The Divine Kingship of the Jukun: A Re-examination of Some Theories." *Africa* 36 (2): 135–52.

INDEX